"A fascinating story, which Zimmer unfolds as a tale of high-stakes scientific sleuthing . . . thanks to marvelously lucid writing."—*Booklist*

"More than just an informative book about macroevolution itself, this is an entertaining history of ideas written with literary flair and technical rigor."—*Publishers Weekly*

"Zimmer is a born storyteller and succeeds in giving us pure pleasure while at the same time teaching us up-to-date science."—Ernst Mayr, Museum of Comparative Zoology, Harvard University

"Zimmer, an honored science journalist . . . leaves life among the fossils agreeably bright." —*The Atlantic Monthly*

"Anyone with an interest in evolution should pick up this book to get on the cutting edge of discovery."—Kevin Padian, Professor and Curator, Department of Integrative Biology and Museum of Paleontology, University of California, Berkeley

"From the first page Carl sets his book apart by diving straight into the most neglected, least understood mystery of all: how wholly new body plans and parts could have been created by natural forces that at first glance would seem to work to destroy innovation. Macroevolution is adaptation without a net. Carl's lucid, often lovely prose is making me finally understand how a species could pull it off without plunging into extinction. He is also very deft at crafting quick-beat narrative out of the lives, inspirations, foibles and occasional dastardliness of the scientists who have pursued this question, both historically and in modern times. I fully expect that *At the Water's Edge* will do for macroevolution what Jon Weiner's *The Beak of the Finch* did for microevolution or David Quammen's *The Song of the Dodo* did for extinction. I'm sure the book is going to really soar."—James Shreeve, author of *The Neandertal Enigma*

"Zimmer is an accomplished popularizer of scientific subjects. This book provides a strong basis for the public understanding of evolutionary patterns and processes"—Robert L. Carroll, McGill University, author of *Vertebrate Paleontology and Evolution*

"This most compelling of evolutionary episodes is told with grace and style. Zimmer's book is a rock hammer blow to those who doubt that evolution is an understandable law of nature."—Peter Ward, University of Washington, author of *The End of Evolution*

"It is wicked, I know, but I have the habit of turning over the corners of pages whenever I chance upon something unexpectedly interesting, exciting or informative. Zimmer's *At the Water's Edge* quickly became the most dog-eared book on my shelves."—Harry Miller, *The Times* (London)

"Zimmer . . . presents an excellent discussion of macroevolution."—*Library Journal*

"The story Zimmer tells is a fascinating one, and not only because it is so skillfully written and so readably presented."—*Audubon* magazine

"Zimmer has done a great job, bringing to life two of the great events in the history of life."—Mark Ridley, Oxford University, author of *Evolution*

"With lively style, fascinating portraits of biologists at work, and splendid command of the evidence, Zimmer tells some of the most gripping stories in the history of life. I recommend it heartily."—Douglas J. Futuyma, State University of New York at Stony Brook

"Richard Dawkins and Stephen Jay Gould, move aside. Carl Zimmer brings evolutionary biology to life more vividly than any author in recent memory. This is a marvelous book by one of our best young science writers. Carl Zimmer has persuaded me that evolutionary biology remains a vital, even thrilling field."—John Horgan, author of *The End of Science*

AT
THE
WATER'S
EDGE

AT
THE
WATER'S
EDGE

*Fish with Fingers, Whales
with Legs, and How Life
Came Ashore but Then
Went Back to Sea*

CARL ZIMMER

*Illustrations
by Carl Buell*

ATRIA PAPERBACK

New York   London   Toronto   Sydney   New Delhi

ATRIA
PAPERBACK

An Imprint of Simon & Schuster, Inc.
1230 Avenue of the Americas
New York, NY 10020

This Atria Paperback edition May 2019

ATRIA PAPERBACK and colophon are registered trademarks
of Simon & Schuster, Inc.

For information about special discounts for bulk purchases, please
contact Simon & Schuster Special Sales at 1-866-506-1949
or business@simonandschuster.com.

The Simon & Schuster Speakers Bureau can bring authors to your live event.
For more information or to book an event, contact the Simon & Schuster Speakers
Bureau at 1-866-248-3049 or visit our website at www.simonspeakers.com.

Designed by Pei Koay

Manufactured in the United States of America
15   17   19   20   18   16   14

The Library of Congress has catalogued
the Touchstone edition as follows:
Zimmer, Carl.
At the water's edge: fish with fingers, whales with legs, and how life came
ashore but then went back to sea/Carl Zimmer.
p. cm.
Includes bibliographical references and index.
1. Macroevolution. I. Title.
QH371.5.Z55   1998
576.8—dc21   97-29331
CIP

ISBN 978-0-684-83490-0
ISBN 978-0-684-85623-0 (pbk)
ISBN 978-1-4767-9974-2 (ebook)

To my family

Taken by leaping madness or by fear
The men jumped to the sea. First Medon's body
Changed color to darkest blue and hunched its back;
Lycabas turned to say, "What monster are you?"
And as he spoke, his nose became a hook,
His mouth grew wide, skin tough, and scales
Ran down his sides, and Libys while he struggled
With leaf-grown oars saw his hands diminish
From claws to fins; another clinging fast
To twisted ropes fell backwards to the sea,
Arms gone and legless, his tail crook'd and pointed
As a third-quarter moon. The creatures lashed
At the ship's side, plunging through spray, now up,
Now down to the sea's floor, swaying like dancers
At a drunken feast, their bodies flashing,
Lips and nostrils pouring spray, they clipped and spawned.

—OVID, *THE METAMORPHOSES* (TRANS. HORACE GREGORY)

Our ancestor was an animal which breathed water, had a swim bladder, a great swimming tail, an imperfect skull, and undoubtedly was an hermaphrodite! Here is a pleasant genealogy for mankind.

—CHARLES DARWIN, LETTER TO THOMAS HUXLEY

CONTENTS

LIFE'S WARPS

Near my left side a yellowtail snapper hovered, breathing water, stammering its fins gently, and flicking its sulfur-striped tail. I was kneeling in the sand at the bottom of the ocean, water piled fifty feet overhead. All I could hear was the static when I sucked air from my scuba regulator and the swarm of my exhaled bubbles as they rose in a confetti column. A long gray shape moved overhead, wheeling and bending, and it took me a moment to recognize it as an Atlantic bottlenose dolphin. The yellowtail snapper and I watched it swim, raising and drawing down its tail a few times, and then gliding, moving like an iron ingot pulled invisibly by a magnetic field. Then suddenly it swept upward, kicking a bit to rise to the wrinkled ceiling of the ocean. When it leaped out of the water its head vanished and then its fins and flukes. For a moment it didn't exist, and then it was drilling down through the water again.

I had come to this place off the coast of Grand Bahama Island to watch how scientists study dolphins. A group of zoologists had piggybacked their research on the dives of a tourist outfit that offered its customers the chance to be entertained by trained dolphins. The customers were ferried out a mile from shore, dove underwater, and formed a circle around two trainers, who carried white drums of dead herring slung on their shoulders—the smell of which had brought the yellowtails here. At the command of the trainers, the dolphins delivered hoops to the customers, pushed against their outstretched arms to wheel them around like turnstiles. The scientists, who wanted to understand how dolphins swim, filmed the animals, and from time to time the trainers would measure the heat flowing from their bodies by pressing a sensor to their flanks. I could sense intelligence, even personality in the dolphins, but their gray masks, their rigid smiles wouldn't reveal how much they enjoyed the process. They seemed to know the rules of this game. If they played along, they got fish; if they decided to break away and explore the water for a while, it was no great loss.

As I knelt there, fish beside me, dolphin overhead, an appreciation of my place in evolution hit me. This was the first time I had dived in the open ocean, and I couldn't stop thinking about how I didn't belong underwater. I needed a steel tank to carry my air, a mask to see, a wetsuit to trap my heat, weights to sink, an inflatable vest to rise, fins to swim. The yellowtail next to me was beautifully designed for living in the ocean: it gulped down water, a squirt of the ocean flowing through its mouth and into its basket of gills, where thin-walled blood vessels traded carbon dioxide and ammonia for oxygen. Flaps over the gills opened, its mouth closed, and the stale water flushed out. Its body had nearly the same buoyancy as water, so that gravity meant little to it. To swim it needed only to flick its body, its knife-shaped profile barely making itself known to the oncoming water. If I reached out toward the yellowtail, it could see my hand, but with lines of pressure-sensitive hairs it could also feel the slush of water that preceded it, and flick away to safety.

Although the dolphins were profoundly different animals from fish, they were manifestly at home in the water as well. They had no gills; to breathe, the dolphins would rise to the surface, open the blowhole at the top of their heads, and suddenly exhale and inhale air. Returning underwater, they would play for a few minutes, all the while holding their breath, kicking their tail up and down rather than side to side like the fish. The dolphins could see the divers about as well as the divers could see them, but they could also visualize the ocean with sound. They emitted high-pitched clicks from their foreheads and listened for the echoes. In their oversized brains the dolphins used the sound to build a picture more precise than their eyes could offer, actually seeing the interior of the fish and humans around them, seeing yellowtails swimming a hundred yards away. The dolphins may have also been communicating to each other with the same kind of sounds, and I wondered what they might be saying. They had been harassing each other the night before in their pen, their flanks covered by raked tooth marks, and so maybe they were trading insults. Maybe they were trying to guess why exactly these humans would give them so much fish for doing these minor rituals.

I can only visit this place. If I were foolish enough to try to stay even an hour here underwater the needle on my tank's gauge would incline to zero and I would pull pointlessly for air. I'd probably panic and flail, my regulator flying out of my mouth, water rushing in. The salt would make me gag, drawing the water down into my lungs. Although there's oxygen in seawater, lungs are unequipped to extract it. Instead, the water tears apart their microscopic pouches braided with blood vessels, makes them swell shut. Unable to unload carbon dioxide into my lungs, my blood would turn vinegarish and my kidneys would burn out trying to neutralize the acid. Meanwhile my circulation would break down, blood sloshing backward in all the wrong directions, my heart beating like a snare drum until it could no longer get the oxygen it needed and stopped altogether. I might try to save myself by speeding to the surface like a dolphin, but my haste would kill me. At the ocean bottom, nitrogen gas dissolves into pressurized blood, but as I rose my veins would bubble like a just-opened bottle of beer. The nitrogen bubbles that formed as I ascended would rove my body, blocking up vessels in my heart, in my brain.

Like all humans, I function best on land. We stand and our weight settles onto our skeletons comfortably, cushioned by pads and sacs. We push our feet down against earth and walk. Our interiors are a set of moist pouches, loaves, and tubes, and our skin does a magnificent job of keeping most of the water from escaping. We draw air into our bodies and extract the oxygen in the nest of alveoli in our lungs, kept from clumping by soapy films, and our blood vessels dump out carbon dioxide in the exhale. We hear ripples in the air millions of times too faint to feel, and create ripples of our own to speak to one another.

The yellowtail's gills can breathe only if they are fanned out underwater like the hair of a swimmer, so that each vessel gets enough room to mingle individually with the chemistry of seawater. If a fish is hauled out of the ocean and dropped in the bottom of a boat, its gills, like the swimmer's hair, fall into a matted clump. Carbon dioxide and ammonia build up in its body, vying for the distinction of poisoning the fish. Its fins and tail, so efficient at pushing water to move its body forward, can only wave to the sun.

If a dolphin beaches, it can survive only a few hours longer than a fish. It still pulls air through its blowhole into its lungs, but its long legless body settles helplessly into the sand. Its overburden of blubber and bulky back muscles crushes down on its lungs and blood vessels. In the water it can carefully manage the heat of its warm-blooded body, but lying on the shore, it can chill or bake depending on the temperature of the air. Before long its entire circulation collapses, veins tear, blood drains downward and pools in its guts. A dolphin's heart can sustain the brain for a time after the rest of its body has shut down, and so it surveys the destruction before it shutters itself as well.

We three animals live in separate countries divided by a fatal boundary. Yet a dissection would show that we are not complete strangers. I volunteer as the human specimen: crack my ribs open and a pair of lungs hangs alongside my esophagus, and they match the pair inside the dolphin. The dolphin and I have giant brains wrinkled with neocortex. We keep the cores of our bodies around ninety-nine degrees. We both fed on mother's milk. And while the dolphin maneuvers with what are called fins, they are actually not like those of the yellowtail. They are in fact camouflaged hands: take away the blubber and gristle and you find five fingers, wrist, elbow, and shoulder.

The similarities between humans and yellowtails are of a more basic sort— we both have skulls and spines, muscles and eyes; we burn oxygen and build our tissue with the hydrocarbons we eat. And some subtler clues reveal that we humans are not the perfect land creatures we might imagine ourselves to be. Look again inside my opened ribs: nestled between my lungs is my heart, and sprouting from it is an aorta that rises upward, sending smaller arteries off toward my head before hooking around and down toward my legs. An engineer presented with a beating heart might have come up with a more rational solution: build two arteries, one to supply blood above the heart, one below.

To understand this tortuous layout, it's necessary to shrink me back down to a seven-week embryo, when my arms and legs were paddles rooted on a body that looked like a pot-bellied, hunchbacked minnow. Just below my eyes and what passes for a brain, six pairs of pouches have formed that look very much like the pouches that house the gills of a fish. My aorta looks far different at this stage: it rises up to my throat and then branches, each branch running through a pouch and then looping back to join another vessel running back down away from my throat. As an embryo, the yellowtail had the same design, but it retained much of it as an adult. The heart pumps oxygen-poor blood into the pouches, which have become bony gills, and then it returns to the vessels that will carry it to the rest of the yellowtail's body. In my case, however, the pouches disappeared into my head and neck as I developed. The cells destined for gill-supporting arches became my voice box and ear bones. The tree of vessels was pared away, a few merging together. And the lowest loop in the primordial set became the hook of my aorta.

The only way to begin to make sense of all these patterns is to understand this meeting underwater—the yellowtail hesitantly hoping to grab some of the dead herring, me kneeling in the sand, the dolphins rolling and flapping—as a family reunion. A little over half a billion years ago the first vertebrate appeared on earth, a jawless, armor-coated, bottom-swimming creature that could fit on your palm. Its descendants diverged into dozens of major lineages that swam through the oceans, including a widespread group called the ray-finned fish, whose twenty-three thousand living species include sturgeons, perch, goldfish, and the yellowtail.

A closely related lineage went through a fabulous shape-shifting: between about 380 and 360 million years ago they evolved legs and feet, lost gills, hooked their aortas, adjusted their bodies in countless other ways, and emerged out of the water to live on dry land. We are descended from these altered fishes, as is every vertebrate that lives on land, be it a skink, a hawk, a tortoise, or a tree frog. We all have our individual names, but we all also share the title *Tetrapod*. The name means "four feet." Among many tetrapods these four feet have gone

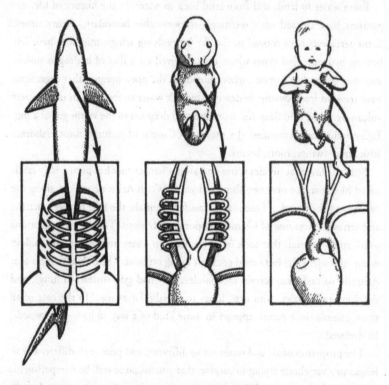

Blood vessels running from a fish's heart branch into its gill pouches. They pick up oxygen there and then join together in a vessel that runs along the fish's back, carrying the blood to the rest of its body. A human embryo forms the same structures, but later pares them down to a form suited for our own air-breathing existence.

through some drastic changes—snakes lost them altogether and birds turned their forelegs into wings, while we stopped walking on our front pair and named them hands. But changed names can't hide the fact that we and all other tetrapods that have overrun dry land owe our existence to one of the most remarkable evolutionary transitions of all. How our ancestors made it through determined much about us humans today.

The reason that dolphins and humans have so many hidden similarities is that dolphins are also tetrapods, not fish. In fact they are mammals like ourselves, with a common ancestor perhaps 80 million years old. The lineage that gave rise to dolphins, whales, and porpoises went through a transformation just as staggering as the one that brought vertebrates on land in the first place: about 50 million years ago a race of wolflike mammals began to adapt themselves to water. They lost their hind limbs altogether and turned their back on the place that had been their ancestors' for over 250 million years.

From water to land, and from land back to water: in the history of life, organisms have crossed such seemingly impenetrable boundaries many times. Some tetrapods have moved to the sky by evolving wings; microbes have left boiling hydrothermal vents where they survived on a diet of hydrogen sulfide and then learned to live on carbon dioxide in the open ocean; other organisms have traveled from swamp to desert, from salt water to fresh, from underwater volcanoes to the cold deep sea, from the cold deep sea to the warm gut of a pig. Every transition demanded the creation of some of nature's most elaborate structures—wings, roots, brains.

Still, the line that divides water and air is a barrier that has particularly fascinated biologists for over two thousand years. When Aristotle walked along the beaches on the island of Lesbos and classified animals, the split between marine and terrestrial was one of his most important divisions. There were exceptions to his rule—animals that took in air but lived in water, and others that took in water (through gills) but could get their living on land. The only offenders that Aristotle named were certain salamanders that had gills instead of lungs, and dolphins and whales, who used lungs to breathe far at sea. "In the case of all these animals their nature appears in some kind of a way to have got warped," he declared.

The properties of air and water are so different, and pose such different challenges to a vertebrate trying to survive, that you might as well be comparing life on two different planets. Water has only a thirtieth of the oxygen found in the same amount of air, and while atmospheric levels of oxygen are almost always steady, they can fluctuate wildly in water. A warm-blooded animal will lose heat to water twenty-four times faster than air—and yet in an entire year water may never experience the swings of temperature that air does in a single day. Sounds race through water four times faster than through air, and it conducts electricity well, but light can only travel a few yards before it fades. Given that water is eight hundred times denser than air, it's far harder to move through in some respects, but if an animal matches its buoyancy to water it is freed from the gravity that tethers animals on land. In the water you want to slip forward. On land you want to fly, if only for a moment.

The place to begin understanding these transformations is with Charles Darwin's *On the Origin of Species.* Evolution, Darwin realized, is always going on around us. Every generation of a species is a lineup of variants, and natural selection, picking and choosing among those variants, allows fitter ones to leave behind more of their offspring. Generation by generation, this skewed reproduction adapts life to its circumstances. Biologists understand this kind of evolution so well that they can now track the success of good genes and understand why not-so-good ones hang on as well. They can predict how droughts will change the beaks of birds; they can measure how long it takes for a moth to adapt to a new pesticide; they can reconstruct the spread of HIV from monkey to man; they can watch evolution happen among the competing cells inside a tumor.

At this generation-by-generation pace, evolution makes teeth thick and tail feathers like lyres. But it also works on longer, grander scales—after all, evolution invented the teeth and the feathers in the first place. To distinguish between these two scales, biologists often used the terms *microevolution* and *macroevolution.* Microevolution has been so well studied that it's changing from a frontier discipline to an applied science, the province of conservation biologists and genetic engineers. But macroevolution has been far harder to tame. There are many things that happen in the history of life beyond the generation-by-generation world of microevolution: continents crash into each other and climates change; dinosaurs go extinct and mammals thrive; new kinds of bodies are built for new kinds of lives.

This book is about the transitions from fish to tetrapod, and from land mammal to whale—two of the most beautiful opportunities for studying macroevolution. Is it, as some have suggested, nothing more than microevolution writ large, or does it call for an expanded conception of evolution? For 140 years the gaps in the fossil record between fish and tetrapods, and between land mammals and whales, were embarrassingly huge, but in the 1980s and the 1990s, a series of discoveries have filled them in. At the same time, other scientists have learned how to look back at these transitions through other evolutionary telescopes—in the comparison of genes, of development from eggs, of locomotion and neurology. Now the twin stories of how vertebrates came ashore and returned to the sea have changed from two of the worst to two of the best documented transitions. There's something telling in the speed of this scientific turnaround. It suggests that macroevolution is on the verge of coming into its own—a rigorous science like microevolution, but in its sweep something far more majestic.

Armed with this knowledge, you may feel a certain kinship with the rest of creation if you happen to find yourself at the ocean floor surrounded by yellowtails and dolphins. At least until your tank runs low.

Chapter One

AFTER A LOST BALLOON

In a basement laboratory in London a man contemplated a carcass. It was roughly the size and shape of a rolling pin, with a knife-edged tail, with fins like whiskers. Its eyes were the dark heads of pins, and its lips were full and sinful. The man stood tall and a little stooped, and although in 1839 he was only thirty-five, he had the kind of brown-eyed glare on which old failed prophets usually have a monopoly. He had seen the skin, guts, and bones of hundreds of animals that no Englishman had seen before, but few had irritated him more than this one.

His name was Richard Owen—not yet Sir Richard, but on his way there. He had grown up in Lancaster, and in boyhood his prospects had not seemed so good. His father, a ruined merchant, had died in the West Indies when Richard was five. A lazy, impudent boy, he joined the navy at fourteen, working as a surgeon's apprentice, but after the end of the War of 1812 peacetime offered little hope of a career, and so he returned to Scotland. He had loved heraldry, and it may have been while drawing unicorns and griffins that he first became interested in animal form. Again employed as a surgeon's apprentice, he began collecting skulls, beginning with dogs, cats, deer, and mice. But he wanted more bones. For a time he worked for a surgeon who performed autopsies on the dead prisoners at Hadrian's Tower. On a cold day in January the two of them paid a call to the prison, and after the autopsy the surgeon left Owen to clean up. This time Owen did not pack all the instruments. He left a few blades alongside the opened corpse, and handing the guard some coins, he told him that he needed to come back later. There was some extra business to attend to before the coffin was screwed shut.

He returned in the frozen moonlight up the icy hill to the tower, bringing with him a strong brown paper bag. He nodded to the guard, who nodded back, and he climbed up the stairs to the autopsy room and shut the door as he went in. Before long he opened the door again, his paper bag full. He passed the guard again and told him that the inmate was ready. As he left the tower and began to walk back down the hill, his mind was filled with thoughts of facial angles and osseous tissues. The more he thought about the work he would enjoy at the surgery that night, the faster he walked. Suddenly his feet kicked out from under him. The bag swung away as he fell on the ice, its contents rolling free. He ran after it, down toward a cottage at the bottom of the hill, but he was too late. A woman inside heard a thump at the base of her front door. She opened it and before her was a disheveled, owl-eyed boy trying to stuff a freshly severed head into a bag. He ran with it all the way to the surgery, her screams flying after him.

Owen went to the University of Edinburgh at age twenty, but he soon outstripped their classes—within a year he was traveling to London with letters of introduction from his private tutors and got a position as a surgeon at a hospital. To be a doctor in London in the 1820s was as much a political act as a profession. The medical establishment—including the city's hospitals and medical journals—was controlled by physicians who generally had Tory sympathies. Allied with the physicians were the surgeons who performed dissections and held a lower position on the social ladder. Opposed to them were the general practitioners who were educated at small private colleges and were of a more radical egalitarian bent. Set on reforming medicine, they lambasted the elitism of physicians and surgeons wherever they could, in renegade medical journals or newspapers or Parliament. They fought fiercely for access to the library and museum of the Royal College of Surgeons. A famous British surgeon named John Hunter had amassed thousands of books and a collection of thirteen thousand pickled animals, disembodied hearts, limbs, bladders, spines, and other materials that laid out the patterns of anatomy, but after his death in 1793, the collection languished for years. The government bought it and entrusted it to the Royal College of Surgeons on the understanding that they would publish catalogs, keep a lecture series, and open the museum and library to the public. But for over twenty years the college had let in only its own, and the GPs crowed loudly until Parliament took notice. The college agreed to let licensed doctors in and to draw up an official catalog.

Owen was hired to assist on the job, and when his superior died, the work became his alone. It was a labor that would keep him occupied for decades. At the same time he struggled to advance his own career as an anatomist, lecturing to medical students and winning the right to dissect the animals that died at the London zoo. It wasn't out of the ordinary to walk into Owen's house and find a freshly dead rhino parked in the hallway.

In 1830 a sixty-one-year-old French baron, ailing yet imperious, came to the Hunterian Museum. Because Owen was the only one there who spoke French, it fell to him to tour the man around. His name was Georges Cuvier—a professor at the National Museum of Natural History in Paris, and to Owen the sun in the sky. Like Owen, Cuvier had started as a shabby-genteel outsider, a German Protestant who wandered the hills around Montbéliard as a boy, picking plants that he would take home. There he would classify them according to the scheme of the eighteenth-century naturalist Carl Linnaeus, who had constructed the hierarchy of species, genus, family, order, class, phylum, and kingdom. Cuvier was supposed to become a bureaucrat in the Prussian government after he graduated, but he emerged out of the Stuttgart Academy among a huge bolus of aspiring bureaucrats. A job came his way only when a friend returned home to Montbéliard after finishing a stint as a tutor to a family in Normandy. He arranged for Cuvier to take his place.

In Normandy Cuvier was treated more like a family servant than a tutor. He tried to console himself by thinking about the tour of Europe he was going to

---

CLASSIFICATION ACCORDING TO LINNAEUS

SPECIES	*Homo sapiens* (modern humans)
GENUS	*Homo* (includes other species such as Neanderthals)
FAMILY	Hominidae (includes hominids such as "Lucy")
ORDER	Primates (includes apes, monkeys, and lemurs)
CLASS	Mammalia
SUBPHYLUM	Vertebrates (includes all animals with bony spines and skulls)
PHYLUM	Chordates (includes all animals with protected spinal cords)
KINGDOM	Animalia

take his student on, and by studying whatever life he could find—the plants in the town garden, a private store of exotic fish, another of oriental birds. But before his tour could begin, the French Revolution overran Paris and spread out into the countryside. His family fled to a village on the coast where Cuvier, now twenty-one, fell into almost complete solitude. He would walk the beach alone, collecting the animals that washed ashore, picking through the guts of skates that the fishermen would sell him. He wrote learned letters to the leading zoologists in Paris, and his reputation as a homegrown naturalist increased. After the revolution cooled, Cuvier visited Paris, where he stunned the scientists at the new National Museum with the knowledge he had taught himself. They were desperate for skilled help and hired him instantly. He was only twenty-three when he joined, but then again the man who invited him, a professor of zoology named Etienne Geoffroy Saint-Hilaire, was only twenty-one. "Come to Paris," Geoffroy wrote to him. "Come play among us the role of another legislator of natural history." It was a parliament founded by boys.

By the age of thirty Cuvier had invented modern paleontology. People had collected fossils for centuries, yet even in the 1500s Europeans thought of them not as real skeletons but as some of the many forms a rock could take: some became emeralds, other became imitations of snail shells. Gradually naturalists recognized too many similarities between these stones and the bones of living animals, mineralogists realized how bone could be transmuted to rock, and the traditional view of fossils began to buckle. By the 1700s naturalists had found bones of everything from elephants to giant spiraled shellfish in the earth. Noah's flood, many of them decided, must have cast them to the tops of mountains.

Cuvier studied fossils as seriously as living animals. Bringing a steady flow of bones to Paris from the limestone quarries outside the city, he prised them free of the rock and found among them elephants that were different from the species alive today. He could only conclude that they belonged to a species that was gone from the world. No one had seriously thought that an entire species could become extinct, but within a few years Cuvier had found rhino-sized sloths and tapirs that had also vanished. He realized that a single flood couldn't have washed away all the creatures he was finding: their fossils disappeared at different points in geological time, demonstrating that life had shuddered over and over again with violent revolutions. Each time a new kind of life had somehow come into existence.

Whenever Cuvier looked at an animal, extinct or alive, he was overwhelmed by how all the parts fit together in a whole—a single, unified organism. A bird was dedicated to flight in every aspect, from the fan of its tail feathers to its enormous lungs to its thin hollow bones. Substitute a bone from a barracuda into a bird skeleton, and the whole creature would be ruined. It was this dedication to function that could account for how different animals were similar to one another and could be fit into Linnaeus's classification. Since function dictated form, animals that functioned similarly looked alike. He divided animals into four major groups, based on how their nervous system—to Cuvier the essence of an animal—was laid out. Vertebrates had a brain and a spinal cord, the mollusks had a brain but no cord, others had barely a nervous system to speak of, and still others had systems that radiated out from a central nerve cluster. With such different architecture, each group could share no connection with the others, and the only reason that animals fell into a given category was that they happened to function in similar ways.

It's no wonder then that Cuvier scorned the ideas of another scientist at the museum named Jean-Baptiste Lamarck, who argued that species actually changed over time—that they evolved. When Lamarck looked at fossil mollusks he found that he could arrange some of them into a smoothly graded sequence through successive layers of rock. He claimed that in coping with a changing environment, a generation of creatures would undergo changes that they would pass on to their own offspring. Cuvier would have none of this transmutation. When Napoleon came home from Egypt with plundered mummies of cats and ibises thousands of years old, Cuvier compared them to living specimens and could find no significant difference. To him, life had a deeper order than Lamarckian evolution could allow.

By the time Cuvier came to London and met young Owen, he had become a hero to France. "Is not Cuvier the greatest poet of our century?" asked Balzac. "Lord Byron reproduces mortal throes in verse, but our immortal naturalist has reconstructed worlds from a whitened bone. Cuvier is a poet by mere numbers. He stirs the void with no artificially magic utterance; he scoops out a fragment of gypsum, discovers a print-mark and cries out 'Behold!'—and lo, the trees are animalized, death becomes life, the world unfolds."

He was admired by the English as well, particularly by the sort of English who ran the College of Surgeons. The agitating general practitioners in London liked the idea of transmutating species and thought that humanity should strive upward, while the upper crust of scientific society preferred the idea that God had created fixed species on earth out of nothing. Their intricate designs were proof of his existence as much as a watch is proof of a watchmaker. And just as every animal had its place, higher or lower, in nature, people had their own place in society. Cuvier was therefore most welcome at the Hunterian Museum, and Owen must have managed to make a good impression on him during their short tour, because Cuvier invited him to Paris the next year.

In Paris Owen spent mornings in Cuvier's collections searching for inspiration for how to organize the mess waiting for him back at the Hunterian Museum. The rest of his time he spent as a dandy, taking in the opera and violoncello lessons, with Saturday nights reserved for Cuvier's soirées—the only time when the stony man would abandon politics and anatomy. Owen also went to meetings of the Academy of Sciences, where people were still buzzing about a debate that Cuvier, after coming back from London, had entered into with his old mentor Geoffroy Saint-Hilaire.

Geoffroy and Cuvier had drifted far apart over the years. Cuvier liked to stick to facts, to avoid the sickened dreams of theory that weren't solidly embedded in the details of real zoology. Geoffroy meanwhile was enchanted by the work of German biologists and philosophers of the day, Romantics who hoped to find a hidden unity to all creation. He had never been much impressed by the classes and orders and kingdoms that scientists like Linnaeus and Cuvier depended on to categorize life. To Geoffroy they seemed arbitrary, since no division was absolute. Things that might make one cluster of animals unique could often be shown to be just the transformed structures of other beasts. A rhino's horn, he showed, was nothing more than a packed clump of hair. And if each species was perfectly created to fit its function, why had God left so many careless mistakes? Why does an ostrich have a wishbone, whose only function in birds is to help them fly?

Geoffroy conceived instead of a wild German hallucination: that you could transform any animal, be it a dog, an ant, or a squid, into any other. The transformation might be painful—to see the underlying similarity between a duck and a squid, you'd have to bend the duck's back into a horseshoe. Eventually Geoffroy decided that these transformations were not hypothetical—they were a sign that evolution had changed old species into new ones. Perhaps, he suggested, the change in an environment changed the way embryos developed. Freaks might be the start of new species. Geoffroy knew that his theory went against everything that Cuvier held dear, and for ten years he tried to lure Cuvier into debate. Finally, after he gave a lecture showing how he could bend the back of a vertebrate and end up with an invertebrate, Cuvier accepted the chal-

lenge. Cuvier had become frustrated by how his own students were being seduced away by the German song of unity, how political revolutionaries were
coming into vogue again, and decided that this was the time to stand against it.

In a series of lectures, Cuvier used forty years of anatomy to mock Geoffroy's
somersaults of bone and muscle. Cuvier was sure afterward that he had won the
debates, but the real results weren't so clear. Owen was a case in point: listening
to biologists argue the opposing views at the academy, he decided that Cuvier
was right, but he still jotted down questions to ask himself. "Unity of Plan or
Final Purpose, as a governing condition of organic development? Series of
species, uninterrupted or broken by intervals? Primary life, by miracle or secondary law?" And within a decade he would try, in his own way, to reconcile the
arguments of these two old Frenchmen.

As Britain's colonial tendrils reached around the world, many of the animals
they encountered came to the Hunterian Museum, and ultimately ended up
under Owen's knife. In the years after he returned home from Paris, he dissected
acouchis, Tibetan bears, kangaroos, tapirs, crocodiles, beavers, mandrills, tou-
cans, cheetahs, hornbills, kinkajous, Indian antelopes, turkey buzzards, water
clams, flamingos, armadillos—both nine-banded and weasel-headed—and parasites from a tiger's stomach. In his work he took special care to stomp out the
dreams of Lamarckian transmutation. In the 1820s biologists had thought that
the bizarre duck-billed platypus was in the same family of mammals as sloths
and anteaters, but in later years stories had streamed from Australia that the animals laid eggs. What could be better proof of a transmuted species? Here was a
living transition between egg-laying reptiles and live-bearing mammals. Geoffroy went on record delighting in the way the platypus bridged these two kinds
of animals.

In 1832, fresh from Paris, Owen confronted Geoffroy over the eggs. The evidence was vague; Owen couldn't find shell-secreting membranes for building
the egg in the womb, nor did the pelvis seem wide enough to let an egg pass
through. It was the ability of a mother to nurse her babies that Owen declared
was the hallmark of all mammals, and he cut open hairless nestling babies to
find coagulated milk in their stomachs and even found that glands in the
mother's belly secreted it. Geoffroy admitted defeat. In fact, the platypus is *both*
a milk-producing mammal and an egg-layer. Its eggs would not be discovered
for years to come, but Owen could have easily noticed a special egg-cutting

tooth in the mouths of platypus babies, just like that in the mouths of reptiles. If he did, he completely ignored it.

In the 1830s English explorers brought back apes to London's Zoological Gardens for the first time. Most of the apes quickly died in their cages, and it was up to Owen to see what they looked like on the inside. Chimpanzees, with their childlike faces, had made a particularly big sensation in London and Paris; Lamarck had even suggested that chimpanzees could have produced the human race. If they should be forced to the ground they would lose their grasping big toes as they became used to walking. Once there they could stand erect, and after generations their calves would develop. With hands free, they would no longer need their jaws for weapons. Their snouts could shorten and their faces flatten, into our own image.

Owen couldn't stand the thought that humans were nothing but upright apes, and even more important, he didn't think much of the science behind it. Geoffroy and other biologists had measured the angles of the faces of orangutans and chimps and had said that they formed a beautiful sequence of flattening that ended with our own flattened visages. But because apes in zoos always died before they matured, the biologists had been limited to studying baby chimpanzees. In 1835 Owen was the first to dissect an adult chimp, and he showed that any resemblance to humans didn't last long: as a chimp grew, its humanlike face bulged out with large, sharp teeth and brow ridges, until its facial angle swung far from our own.

While Owen was busy preserving mankind by the slope of its brow, a naturalist named Johann Natterer was thrashing through the Amazon. Among the hundreds of animals he trapped, one that he found swimming in a river gave him particular pause. He was so confused by it, in fact, that he brought his specimens to Leopold Fitzinger, the curator of reptiles at the Imperial Museum in Vienna. They looked like fish, with gills and a fin, but when Fitzinger probed down their throats, he found what looked, absurdly, like the traces of a lung. Was this a fish or a reptile (a term that in Fitzinger's day included amphibians like frogs and newts)? No one had ever been forced to make such a distinction before—after all, what distinction could be clearer? The only animals that had dared approach this division before were eel-shaped salamanders called sirens, which often lived underwater, breathing water with a feathery set of gill-like structures. Yet even they had legs and toes. Fitzinger settled on the title of *reptile* but only lightly, given how badly gutted his specimens had been—"victims," in his words, "of Natterer's too passionately executed chase." Natterer

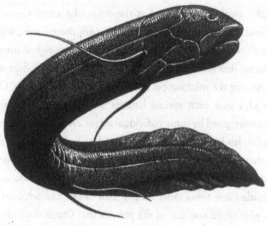

Lepidosiren paradoxa

himself had thought it was a fish, but he bowed to Fitzinger's expertise. He named the creature *Lepidosiren paradoxa: lepido* for the scales on its back, *siren* for the amphibians, and *paradoxa* for his confusion.

In June 1837 another specimen came to the Royal College of Surgeons, but instead of the Amazon, this creature lived in the Gambia River in West Africa. During the dry season there, the creature dug itself a burrow a foot and a half deep in the mud, where it stayed for months. Owen's specimen came locked in a vault of clay. Inside was the same rolling-pin body, the lips, the whiskerlike legs that Natterer had found. But Owen hadn't yet heard about the reptile from the Amazon, and so he promptly gave his own creature a name: *Protopterus anguilliform*.

Owen set the creature aside. If it had been a few years earlier, he would have given it his relentless attention, but fame and work were swallowing up his days. The Royal College of Surgeons tore down the old Hunterian Museum and built a bigger one in its place, with collections finer than Cuvier's. Owen won awards and gave lectures that London's high society now came to hear. And on top of all this came tons of new fossils, brought to Owen by a quiet, pug-faced naturalist names Charles Darwin. Darwin had gone to the University of Edinburgh a year after Owen, but the two of them met for the first time only in 1836, at the house of a mutual friend. There Darwin told Owen about the five-year trip around the world he had just completed and the treasures, such as hideous green lizards that leaped into the Pacific to eat seaweed, that he had brought home. He had also discovered giant mammal fossils from South America he couldn't make much sense of, and at the dinner he asked Owen to identify them. It would take years for Owen to finish with these bones, which he concluded were rhino-sized rodents and anteaters that outweighed horses.

Often during that time Darwin would visit Owen at the college to talk about the fossils. Among the microscopes and jars of preserved animals, Owen would explain his idea that each species had its own organizing energy, one that couldn't be overstepped by some individual intent on founding a new one. Darwin kept quiet, his head steaming with new ideas he was afraid of sharing with anyone. Perhaps as they spoke Owen resented this lucky, quiet man. Here was Darwin, who had been born with enough family wealth to do as he pleased, who now made three times more money than Owen had achieved with all his struggling, who could stay out of the politics that Owen depended on for his livelihood. Owen was a museum man, but Darwin could afford to go on long voyages, to spend his days back in England doing little more than raising pigeons or chatting with dog breeders or strolling to the zoo to watch an orangutan as it threw tantrums.

Owen meanwhile was so submerged in work that he discovered that for once someone had beaten him. The African creature he wanted to call *Protopterus* had already been found in the Amazon and already named *Lepidosiren*. Zoologists treasure the opportunity to name a species, as if it were handed down to them by Adam himself, and Owen was no exception. If anything, he felt more of a right than his peers. Owen was the English Cuvier, and he knew it—he had become an arrogant, secretive, manipulative man—and in his writing about *Lepidosiren* you can sense his resentment that such an important animal should be left to anyone else to name. "Since the time of *Ornithorhynchus*," he wrote (referring to the platypus), "there had not been submitted to naturalists a species which proved more strongly the necessity of a knowledge of its whole organization, both external and internal, in order to arrive at a correct view of its real nature and affinities than did the *Lepidosiren*."

Owen decided to judge for himself what this animal was. In his basement laboratory he looked at its mouth, which held the strangest teeth he had ever seen—two giant, ribbed crushing plates cemented to its palate. These alone would have made it an exceptional creature: fossils of an identical shape had been drifting through the scientific literature for over twenty years. They were gorgeous black polished things, which their first describer thought were broken off the end of a turtle's shell. Eventually they were recognized as being teeth belonging to a fish that lived during the Triassic period, which geologists would identify as being over 220 million years ago. And here the same teeth were in this living animal's head.

"If indeed, the species had been known only by its skeleton," Owen wrote, "no one could have hesitated in referring it to the class of Fishes." But unlike the fossils, this species had dried flesh attached to it, and that made matters confusing. Owen slit open the dried, olive corpse, picking through the green bones. He laid out the gills from their bony struts so that they hung like mimosa leaves. Within its ribs, where some fish gave a gas-filled sac known as a swim bladder for controlling their buoyancy, *Lepidosiren* had long honeycombed bags that tangled into the heart. He could only call them lungs—"for I know not how otherwise to designate, according to their physiological or morphological relations, those organs, which, in the technical language of ichthyologists would be termed the swim or air bladder." These lungs could only be for breathing air, and once again a transformationist might say that here was a transitional animal bridging the gap from fish to reptile.

Because Owen did not believe in such transitions, he took particular care to figure out on which side of the border dividing fish and tetrapods the lungfish belonged. The brain resembled those of reptiles, and its wispy fins fit into its shoulders and hips like simplified legs. Lungs were surely the most tetrapodlike trait of *Lepidosiren*, the organs that let them survive on land, as opposed to gills which let fish breathe water. But Owen didn't want to reach such a conclusion, and he kept looking at the animal until he inspected the nose. In his specimen there was no passage connecting it to the mouth, which meant that *Lepidosiren* could only smell with it, as did all fishes. There was no way that it could be used for breathing, as in tetrapods on land. Owen had found the sign he needed.

"In the organ of smell, we have at last a character which is absolute in reference to the distinction of Fishes from Reptiles," he declared. "In every Fish it is a shut sac communicating only with the external surface; in every Reptile it is a canal with both an internal and an external opening. According to this test, *Lepidosiren* is a Fish . . . not by its gills, not by its air bladders, not by its spiral intestine, not by its unossified skeleton, nor its extremities nor its skin nor its eyes nor its ears, but simply by its nose."

Owen had wanted to save humanity from the company of beasts. Now we were more than fishes, thanks only to our noble nostrils.

For twenty years after his encounter with *Lepidosiren*, Owen became even grander. He invented the word *dinosaur*, and he held a dinner at the Crystal

Palace at which a table of luminaries sat inside the gut of a model of *Iguanodon*. He was knighted and moved from the Hunterian Museum to the newly built British Museum, where he superintended the natural history collections. William Gladstone and Charles Dickens were among his friends. Queen Victoria gave him a house and he tutored her children.

And in that time Owen found a way to fuse the ideas of Cuvier and Geoffroy. Cuvier had insisted that function determined form, that there was no correspondence between the major groups of animals. But Owen decided that Geoffroy, who could see ways of turning any animal into any other, might have touched the core of life, even if his details were wrong. There was simply too much anatomy that Owen had held in his hands to be accounted for by getting this or that job done. You could talk about how the heads of mammals like ourselves are made of many bones that don't fuse until after birth, and how God had so providently arranged this so that our heads could slip out of narrow wombs. But why then could Owen find corresponding bones in a chicken or a lizard, which only had to break its way out of an egg to be born? There had to be a deeper relationship, which Owen called homology—"the same organ in different animals under every variety of form and function" was his definition. Homology was what united the bones in the heads of birds and humans, but homology always had to be distinguished from analogy when different organs served the same purpose. Gliding lizards have long spars attached to each of their ribs, on which they hang a fleshy sail for parachuting from tree to tree. Gliding squirrels do the same thing with flaps of skin that stretch from arms to legs. Both animals can sail, but the anatomical parts that let them do it aren't homologies.

When Owen looked at life this way, he could see homologies throughout the anatomies of all vertebrate animals. Ultimately, he decided, their bodies could all be considered elaborated spines. The basic vertebra was a spool surrounded by an arch above and below, and ribs protruding from the side. Every part of the vertebrate skeleton was a homology of some part of this bone, from chins to braincases to breastbones. Reducing all of these homologous bones to their simplest form, Owen drew what he imagined was the general plan for vertebrates, a lampreylike thing which he christened the Archetype.

The Archetype was the blueprint that God referred to as He guided the history of life. As Cuvier had suggested, life passed through revolutions, and as it did, different modifications on the Archetype emerged—first fishes, then reptiles, and then mammals. The only continuity they shared was in how they elaborated more and more on the Archetype. The lungfish, the animal that had puzzled Owen so much before, was actually the closest approach that fish made to the reptile form. Although Owen thought that new species appeared through time, he didn't accept Lamarck's idea of life as continual upward progression. To him the dinosaurs were clear proof of that: although they were reptiles, they were advanced creatures in their own right, and yet they had been followed by far less inspiring lizards, snakes, and other minor reptiles.

Some might think that if God had not made animals perfectly adapted to their own ways of life then there could be no purpose to their anatomy. Perhaps they had formed by organic chance. "But from this Epicurean slough of despond every healthy mind naturally recoils," Owen wrote. The Archetype did have a function of its own, he decided, but something more noble than letting an animal trot. God must have had our own form in mind when He laid down the earliest vertebrate forms, and thus "its truer comprehension leads rational and responsible beings to a better conception of their own origin and Creator."

Beneath Owen's pieties, he was secretly trying to figure out how species were in fact created organically. Churchmen might think that all the species alive today came from Noah's ark, but Owen couldn't help wondering how a flightless bird could get to a remote island in the Pacific. He would always consider Lamarck's strivings and Geoffroy's freaks to be laughable, but evolution might work by other ways. Populations of aphids would change completely from one generation to the next, all completely lacking wings, all turned asexual. Whatever rules of biology produced these changes might on rare occasion create new species at a stroke. Of course God was behind whatever mechanism created species, but science could discover it. Owen was always coy about these ideas of evolution. He had clawed for three decades to reach his social and scientific perch, and if he began to offer this sort of heresy in public, he might tumble from it. His secondary laws he simply called "creative acts."

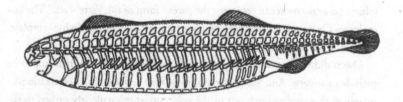

Richard Owen's vertebrate Archetype

In 1859 coyness became obsolete: after twenty years of quiet puttering, Darwin published *On the Origin of Species by Means of Natural Selection.* Its genesis had come in July 1837, only a month after Richard Owen first laid eyes on *Lepidosiren,* when Darwin had opened a red notebook and begun sketching out his ideas for how species could be born. Although raising pigeons might have seemed an odd hobby for Darwin, they were his muses. He could make the birds change their anatomy dramatically in only a few generations simply by breeding only those birds that came closest to his specifications. Nature, Darwin had realized, was a breeder as well, although one without a conscious purpose. An animal faced horrendous odds in its life—disease, predators, droughts, and the competition of the other members of its species. In every generation the individuals all vary in size, strength, and every other trait. Some of those traits would give an animal a slight reproductive edge over others. Admittedly the variations were small, but Darwin pointed out that the earth was much older than people had once thought, and so there was plenty of time for natural selection to work on life. If populations of a species became isolated and faced new challenges, natural selection would push them away from their ancestral form and into a new species. Despite what Owen might say, the fossil record showed that species didn't have any internal energy—a certain species might last far longer than others, thanks only to its ability to outcompete other forms, Darwin decided. Life didn't climb a pole toward our own glorious humanity; it branched off into new species that branched off into newer ones. Some of the branches reached all the way up through time to today, while others had been clipped by extinction.

Darwin's theory was powerful because it lacked the magic tricks of the old Lamarckians—the inner striving, the passing on of characteristics acquired in life—and because he could sweep up the ideas of others and fit them so elegantly into his own frame. He plundered Owen most shamelessly of all. "I look at Owen's Archetypes as more than ideal," he wrote in the margin of one of Owen's books, "as a real representation as far as the most consummate skill & loftiest generalization can represent the parent form of the Vertebrata." The homologies Owen saw among vertebrates were not traces of the divine template, but family resemblances.

Owen didn't appreciate Darwin's twist: an English Cuvier ought not to be prelude to anyone. And when Darwin wrote how certain unnamed creationists imagined that species lurch out of the void, atoms miraculously turned to tissue, he was sure that Darwin was caricaturing himself. Making matters worse was the way that Darwin's champions—people like the paleontologist Thomas Huxley—made Owen the straw man for their evolutionary torches, misreading his objections to Darwin. Owen had become willing to accept that homologies were the result of inheritance, and even that humans might well evolve in the future into a new species altogether. But theories such as Darwin's that tried to explain how evolution could happen were nothing more than "guess-endeavors." Darwin had shown only that natural selection *might* explain life, not that it must. Worse yet, it was an ugly mechanism based on nothing more than random dyings and birthings, not some fundamental law.

Shy and sick, Darwin stayed at his house in the countryside, receiving news of the fate of his idea in London's scientific circles as he raised his orchids and carnivorous sundews. Owen's and Darwin's followers scuffled on the academic platforms for the next fifteen years. It was a bitter fight, full of misreadings and insults that left most parties alienated. "I used to be ashamed of hating him so much," Darwin once wrote of Owen in a letter, "but now I will carefully cherish my hatred & contempt to the last day of my life."

Owen worked against Darwin by continuing to try to separate humanity from the apes. Gorillas had recently come into English possession, and after dissecting them Owen declared that human brains had structures that not even these apes possessed. Since it was in the brain that the lofty mind of man resided, it made sense that the mark of his uniqueness should be found there as well. But in a series of public humiliations, Huxley displayed the results of new dissections of primates, and he showed that Owen had gotten his anatomy badly confused.

To many opponents of Darwin such as Owen, the central horror of his theory was that humans are descended from apes. Yet the transition from apelike ancestors to humans was a late, minor change in our kaleidoscopic descent. At least an ape can walk and breathe air. At least it has hair and thumbs. For real alienation, go back to a fish. Who can see a kindred spirit in those flat button eyes? The flattened or elongated body, nothing more than a mouth driven forward by muscle? Darwin knew this well. He once wrote to a friend, "*Our* ancestor was an animal which breathed water, had a swim bladder, a great swimming tail, an imperfect skull, and undoubtedly was an hermaphrodite! Here is a pleasant genealogy for mankind."

Repugnance aside, Darwin's opponents could have waved *Lepidosiren* over their heads to rally their forces. Getting a fish on land calls for inventing a radically new anatomy to adapt to a place that would normally kill it. How could the little changes of size and shape create complex structures altogether unprecedented, like legs and feet? And where were the intermediates, either alive or among fossils? Darwin had agonized over this problem himself, and in print. How could evolution, he asked, create something as complex as an eye? He could only suggest a pathway—that a patch of light-sensitive cells had curled around into a cup and developed a lens. As evidence, he pointed out that all of the stages of this evolution actually existed and functioned in different animals, from flies to squid to humans.

Here was a new kind of argument. A physicist can observe things like orbiting moons and falling cannonballs, find formulas to predict how any object is affected by a force such as gravity, and run experiments to test them. Before Darwin people thought of biology as a medical philosophy, a way to use bone and flesh to contemplate abstract concepts of classifications and connections; or a kind of zoological Christianity, in which all of the adaptations that organisms made to their surroundings were proof of God's omnipotence. Now Darwin was making biology an historical science. A living bird had its wings today thanks to an inconceivably long, twisted path of evolution that no one could ever hope to reduce to a formula or fully duplicate in some experiment. Nor could we find many transitional species alive to show that history, Darwin warned, because a successful new species would be likely to compete its parent species into oblivion, not to mention the various catastrophes that could wipe out a species. As for fossils, while the earth was a vast museum of bones, its collection was haphazardly chosen. We could only hope to find a few twigs of the tree of life.

When Darwin died in 1882, Darwinism was ailing. A growing number of scientists were accepting evolution, but his version of it stood on wobbly legs. Gaps between different forms of life were still often vast, and Darwin's idea of heredity—a sort of blending—couldn't explain heredity as it actually took place. When Owen died ten years later, he may well have gone to the grave sure that if his own idea of evolution had lost, at least Darwin's hadn't won.

It would be many decades before natural selection on a generation-by-generation scale—what we now call microevolution—was vindicated, and decades more before we began to understand heredity on the level of individual molecules. Deoxyribonucleic acid, better known as DNA, is a double-helix strand that is carried in every living cell of every organism except some viruses.

It is composed of a twisted backbone of sugar molecules and phosphate that, if untangled, would stretch a yard long. Running between them are 3 billion rungs of information—pairs of interlocking molecules known as nucleotides. They come in only four different forms, and serve as a brief alphabet for life's instructions. Floating close by the DNA are proteins that keep the strands wound up tightly and yet allow the nucleotides to be read easily. When signaling molecules arrive from the outside, other proteins lock onto stretches of the DNA and reproduce them in short strands. These copies are trundled to blob-shaped factories floating throughout the cell. At these sites still other proteins use this message as a template for hooking together little molecules called amino acids into proteins. Proteins are the bulk and labor of life: they become hair and skin and connective tissue and fingernails; they are also ferries for oxygen in the blood, digestors in the bowels, light-catchers in the eyes, and handlers of DNA itself.

Life reproduces itself in many ways, but they all involve creating new DNA. In a woman, for example, all cells have two sets of genes that are almost identical, except the cells in her ovaries that ultimately become her eggs. When the time comes for one of these cells to take on its identity as an egg, its genes drape over each other, exchange sequences, and then split apart. The cell divides into eggs, and each one thus has only one set of genes instead of the normal two. A man's sperm likewise has one set of genes, and when it swims up to an egg and fuses with it, their genes combine into a full supply. A baby's new genes are not a blurring of mother and father but a mosaic made from each set. If it gets a particular gene from its mother, it will make the same protein that the gene made in her body. Heredity is the result of this mixing of intact genes rather than the blended frappé that Darwin imagined. Some single genes have great individual power, and they alone can give a baby traits like blue eyes. Others are subtle: they regulate the effects of other proteins, or cooperate with dozens of proteins on some task, such as making cell membranes sticky or slick. In these cases heredity is far harder to tease out.

It is a beautiful system but far from perfect, and its imperfection lets evolution unfold. When sperm and eggs form and when their genes mingle, mistakes can creep in. A gene may get clipped short as it's being traded from chromosome to chromosome, or a base pair may be misread. A glitch in the code often makes no difference, but sometimes it may jam up a gene so that it cannot make a protein, and for want of that protein, bones turn brittle or blood thin, or babies are never born. On rarer occasions it can make a gene that does a better job than the older version, or a new job altogether. If that improvement gives an animal slightly better odds of surviving to adulthood, of mating, of raising its own offspring, it will spread through a population. If it is far superior, the new gene may become universal in a few generations; if it is only moderately better and another variant of the gene crops up frequently, it may simply dominate the population. Some mediocre genes are linked to good ones and are thus inherited along with them. Others can only work well if they're inherited as a group. Sometimes genes simply rise and fall randomly. In a big population this drift is usually only a minor flutter in the statistics, but in a small, isolated cluster of animals, a gene may end up swamping them for no particular evolutionary reason.

Once biologists began to understand how genes mutate, they could finally grasp what had eluded Darwin about the way species actually originate. To make new species, they have found, you need to break down an old one. Say, for example, that a new stream splits up the crickets that live in a valley. They can no longer reach each other to mate, and so genes are circulated only within each group. As the genes mutate, as fitter forms appear or as they simply drift, the crickets become less and less alike. If the stream dries up allowing the two populations to mingle again after a few thousand years, the crickets may no longer be able to mate at all because their genes have become so incompatible.

This is the simplest way biologists know to start a species, but there are many other possibilities, some with more evidence to back them up than others. Animals may initially become incompatible for nongenetic reasons: female crickets choose their mates by listening to their song, and the songs of males on each side of the stream might drift apart so much that the females on the other side find it unromantic. Speciation may happen quickly if the barrier breaks off only a fragment of the population. The gene pool becomes so uneven in a small group of animals that new gene combinations can sweep through them. On the other hand, some researchers suspect that speciation can occur without any split at all. As a species of cricket colonizes a region, the ones at the northern end may become able to withstand much colder winters than their southern cousins. If they mate with the less cold-adapted crickets, their offspring would be vulnerable to the winters, and so they would tend to stick to their own kind. Living side by side, they would become isolated and turn into two separate species.

This combination of Darwin's ideas with twentieth-century genetics is known as neodarwinism. It describes the way that genes evolve, spread, and form new species so well that it is almost workaday. When biologists first proposed this explanation for microevolution, they showed that if you let it run for thousands or millions of generations, it *could* produce the changes in the fossil record. But in order for humanity to witness life's actual history, macroevolution had to become a science of its own. The first step was to figure out how organisms—both living and long since turned to rock—are related.

When later biologists went back to the animals that Richard Owen had studied, they sometimes found mistakes in his work. Some were the result of a bad specimen; in others he simply chose to ignore some clue if it clashed with his own theories. As more specimens of *Lepidosiren* were studied it became clear, for example, that despite Owen's claims they *did* have nostrils—or at least nostril-like passageways—that connected into their mouths. Owen's dried specimen must have had a stuffed nose. Later scientists like the Irish anatomist Robert M'Donnel were left to agonize over their true identity. "It may be presumed that the *Lepidosiren* is invested with a peculiar interest in consequence of its, as it were, standing upon the boundary line between two great compartments of animal creation," he wrote in 1860. "It is in itself a proof of how almost imperceptible are the transitions from one class to another, as we ascend in the great scale of nature."

M'Donnel had received an animal alive in its mud case from near Macarthy's Island on the Gambia River; it had been wrapped in a piece of sailcloth and packed in a box for seventy-six days. "On receiving it I was, of course, very anxious to know whether it was alive. I accordingly having opened the box, pushed a straw into the air hole as to touch it, whereupon it squeaked so loudly, as not only to give the unmistakable evidence of its existence, but to make me quickly draw back my hand, in fear lest I might be bitten." He proceeded to open its cocoon up; the mud was so hard he had to saw it in half. "During this process, the animal in reply produced vocal sounds, unquestionably voluntary, no doubt less musical than the fabled syren-song, but to my ears very agreeable." Inside was the fish, covered in a slough like dried birch leaves. He put it in a tank of water and watched it unroll itself and swim around vigorously. It never sang to him again.

M'Donnell found that the nostrils in this animal were clearly crucial for letting it breathe when it was burrowed in its mud sac. In his opinion its heart was like a tadpole's, its lungs like a snake's. "I know of no animal more calculated leading to the adoption of the theory of Darwin, than the *Lepidosiren*." In other words, he didn't know what to call it.

A third continent, Australia, offered up a cousin to *Lepidosiren* in 1870. This one was pale and broad, and like its relatives in Africa and South America, it had fins that anchored at the shoulders and hips like limbs. Unlike *Lepidosiren*, which had slender, whiplike fins, it had stout ones that contained branches of bone sprouting along a main axis. The Australian species, which rarely came out of water or breathed air, looked even more like an ancient fossil than the others; it dated back to the Devonian period, which lasted from 410 to 355 million years ago, when the earliest fishes appeared. And because of their deep genealogy, this lineage, collectively called lungfishes (or dipnoans, meaning "two lungs"), were pulled into the broader debate over evolution. As one skeptic wrote, "the supposition of some zoologists, who saw in *Lepidosiren* an instance of the latest step of advance attained by the struggling ichthyic type towards the higher class, that of Amphibians, is not confirmed; for we find that the Dipnoans reach back, with comparatively insignificant modifications, into one of the oldest epochs from which fish-remains are preserved." Yet Darwin never claimed that lungfish had given rise to our tetrapod ancestors. Instead, lungfish and tetrapods shared a close common ancestor in aquatic form, and the two lineages had taken different courses through history.

By the end of the nineteenth century paleontologists had gotten a much clearer view of the early history of vertebrates. The most primitive vertebrates—represented today by hagfish and lampreys—lacked jaws and could only snuffle through underwater muck for their prey. Larger forms with jaws evolved about 450 million years ago, and the years that followed were a golden age for fishes. During the Devonian period, a diversity of vertebrates that's never been seen since filled the ocean: sharks with fins covered in teeth, armored placoderms, thirty-foot arthrodires, antiarchs with fins that looked like crab legs, and a riot of giant lungfishes.

While most of these lineages have disappeared, four major groups that got their start in the Devonian are still alive, including the jawless fishes, sharks, and ray-finned fishes. Lungfishes are among the survivors of the fourth group, known as the lobe-finned fishes. They earned their name because, among other things they held in common, they all had fleshy fins with a few sturdy bones inside. And as paleontologists found fossils of other lobe-fins from the Devonian, they concluded that some might be even closer to the origin of tetrapods than lungfish. Lungfish limbs are certainly more like our own than a sturgeon's, but they also have some crucial differences. The shoulder joint of humans and all other tetrapods is a ball and socket with the top of the humerus the ball and the shoulder the socket. In lungfish, however, it's the shoulder that has the ball and

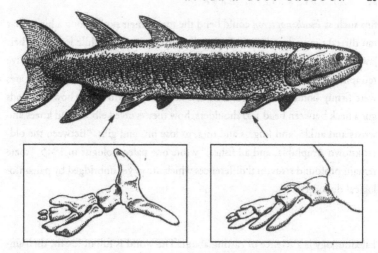

The lobe-fin *Eusthenopteron*. Some of the bones inside its front paired fins correspond to the humerus, radius, and ulna of our arms (right), while some in its rear pair correspond to the femur, tibia, and fibula of our legs (left).

the humerus the socket. Lungfishes have nostrils like tetrapods, but the bones that roof their skulls are great expanses of little prisms that look as if someone hit their heads with a mallet and then glued the pieces back together.

For closer kin to ourselves paleontologists looked to newly discovered lobe-finned fish, such as one genus called *Eusthenopteron*. In the 1880s a farmer in Canada uncovered splendidly unsquashed *Eusthenopteron* fossils, and even better ones have turned up every few decades since. It might not have attracted Natterer's attention if he happened on it in a river; it could pass for an ordinary fish, with a body like a pike or a muskellunge, only belied by a squat head and stout lobes for fins. It controlled them with a set of small bones that were strikingly similar to our arms and legs. The bone closest to your shoulder is the humerus, and the two thin bones below the elbow are the radius and ulna; *Eusthenopteron* had compressed versions of the same combination of bones in its lobes. And rather than a smashed blur, its skull was a series of paired bones much like the ones that run from your nose to the back of your head. When the paleontologists compared bone with bone, noting how they abutted each other, they realized that they were homologies of the skull bones of the early amphibians.

These homologies put lobe-fins such as *Eusthenopteron* close to the origin of tetrapods, but they weren't enough to reveal the full transformation. Many lobe-fins such as *Eusthenopteron* could bend the top of their skulls along a hinge that ran directly over their brains. By bending up their snout while lowering their jaw they could attack a fish with a bigger bite. Even the most primitive tetrapods couldn't do this trick because the bones that made up their braincases were firmly bonded together. And then there were matters like how tetrapods got a neck between head and shoulders, how they evolved elbows and knees and wrists and ankles and fingers and toes, to lose fins and gills. "Between the oldest known Amphibia, and all fishes," wrote one paleontologist in 1915, "there remain profound structural differences which are as yet unbridged by paleontological discovery."

Paleontology is a science of casting about. The world is full of fossils, their unbelievable bodies frozen in a dying twitch or torn apart in the earth's crust, but the world is too wide and too covered over in mud and trees for us to hope to find many of them. Paleontologists need patience, whether they spend an afternoon walking across a desert floor or generations waiting for the discovery of a fossil to bring sense to fossils already found. By the early 1900s, it might have seemed that the earth had nothing more to say about how vertebrates first walked onto dry land. More Devonian lobe-fins came out of their tombs, but the oldest tetrapods dated back to the later Carboniferous era, separated by 60 million years. It was possible that the first tetrapod had appeared in the Devonian, but a century of fossil collecting in the Devonian rocks of England and elsewhere hadn't turned one up.

A doomed balloon ride across the Arctic was the salvation that paleontology needed. In the 1890s European explorers still had not found the North Pole and were resorting to desperate means. Fridtjof Nansen of Sweden sailed directly into the congealing autumn Arctic so that his ship, the *Fram*, would be bound in ice and carried by the ocean currents toward the pole. For two and a half years he and his crew drifted with the pack, until it became clear that the *Fram* had stopped moving north and was moving east toward Europe. He leaped out of the ship and tried to sled up to the pole, only to discover that the ice he was now traveling on was moving south. Only four degrees away from true north, he bolted back for Franz Josef Land.

The *Fram* drifted east for months before it broke free and the crew could sail south to the island of Spitsbergen. There on the bare flats they saw a giant balloon. Its pilot was a young Swedish engineer named Salamon Andrée. Andrée had decided that ships like the *Fram* could never reach the pole, and that flight offered the only hope. He had convinced the king of Sweden and Alfred Nobel to pay for a balloon, which he had brought by ship to Spitsbergen. There he mixed tons of sulfuric acid and zinc to create hydrogen gas, which filled his silk canopy for four days. But gales hit the island before he was ready to launch the balloon, and then the *Fram* arrived with stories of how Nansen was racing on sleds toward the pole. Andrée let the canopy fall back to the ground.

When he got back to Sweden and discovered that Nansen had actually failed, he began to plot a second attempt. He returned to Spitsbergen in 1897 and this time he flew, but he did not need much time to realize that he would fail as well. For a few days he floated north with his crew of two, bobbing up and down with the sudden changes in temperature and moisture of the Arctic atmosphere. As he crossed over the edge of the polar ice the balloon became weighed down by rain and snow, until the guidelines dragged across the ice, until the gondola bounced like a ball on the ground, until the balloon came to rest. For a week the crew huddled in cramped fog. Andrée decided to pack sledges with food and a collapsible boat, which they dragged over the drifting ice. Hauling the sledges across sloshing leads they hoped that they could find refuge in Franz Josef Land. But the ice wandered in the wrong direction under their feet, and after two months of this polar treadmill they reached a little hump of Arctic rock called White Island. Thirty-seven years later whalers came to the island and discovered their decrepit boat, their journals, and Andrée's corpse still sitting in the snow.

But in 1897, no one knew where Andrée had gone. His fellow Swedish scientists searched for him by ship in the following summers, first traveling around Spitsbergen and then heading to Greenland. As the pack ice opened, they traveled for eight weeks along its eastern edge in their sail- and steam-powered ship. They mapped the tentacled coast, and in one fjord along an elephant-backed mountain they named Celsius Berg, the explorers found bones. They weren't the bones of Andrée and his crew, however; they were the bones of lobe-fins, hundreds of millions of years old.

These fossils had been found elsewhere in Devonian rocks, and to those who studied that era, it was as if a new continent suddenly appeared on the map: other Devonian rocks were hidden for the most part under a woody, bushy carpet in places like England and Pennsylvania, while the mountains of Greenland were mercilessly bare. Unfortunately the new fossils were also so remote that only some greater pretext—like the search for a famous explorer—could get the paleontologists to this far corner of the Arctic. Another rationale came about thirty years later in the late 1920s, when Denmark and Norway began competing for control of eastern Greenland and the oil and minerals that it might hold. The Danes brought Swedish scientists with them, and they found more bones belonging to lobe-fins, as well as a few things they didn't know what to make of, simply marking them as "scales of a fish-like vertebrate of uncertain affinities."

These expeditions were only a little less brutal than Andrée's and Nansen's trips. The scientists still traveled in wooden steamers with three square-rigged masts, and while they could now bring a hydroplane for their surveys, they still wore polar bear suits when they flew. In 1931 an energetic twenty-two-year-old geologist named Gunnar Säve-Söderbergh was put in charge of the expeditions. For sixteen hours a day he could climb mountains, throwing rocks into his rucksack and sketching out stratigraphy along the way. He had a book of numbered tags made for the expeditions, *P.* for fishes and *A.* for amphibians—a supremely confident system, considering that no one had ever found a Devonian amphibian. That first summer, as Säve-Söderbergh made his way around the northern slope of Celsius Berg, he found more fossils. In the cones of fallen rocks below the mountain's eastern plateau, he also found more than a dozen scraps of a flat skull that didn't look like any lobe-fin he had seen before. Optimistically, he marked them with *A.* tags.

Back in Stockholm that fall, he slowly worked the bones free of the hard sandstone, using alcohol and balsam to reveal the sutures between the bones. Looking down on the flat roof of the skull, he could see that although some of the bones that made it up were patterned like those on a lobe-fin, it had a long snout and other features that only early tetrapods had. He realized that he had found the earliest tetrapod, and he named it *Ichthyostega*—"fish plate"—after the top of the animal's skull. The discovery was a great sensation in Denmark, not only with the politicians who wanted to tighten their grip on Greenland, but with the public as well. In celebration one newspaper cartoonist drew a trout with dog legs carrying a pipe-smoking caveman, while snakes encircled mountain peaks and elephants flapped their wings overhead.

Säve-Söderbergh spent the following few summers mapping more of the region by foot, boat, and Icelandic horse. Fossils practically fell out of the rocks for him—mostly fish but on rare occasions another piece of *Ichthyostega*. The strange scales that had been found in 1929 turned out to be *Ichthyostega*'s ribs, massive and overlapping like venetian blinds of bone. His assistants, particularly a student from the University of Uppsala named Erik Jarvik, found more *Ichthyostega* skulls. One unearthed in 1934 was so handsome that the paleontologists brought it back across the Atlantic resting on a blue velvet pillow. After five years of these successes Säve-Söderbergh was appointed a professor at the University of Uppsala, but in that year he was also diagnosed with tuberculosis. He lingered in bed, managing to write a few papers about some of the lobe-fins he had collected, and died in June 1948 at age 40. In the summer of Säve-Söderbergh's death, the expedition to Greenland finally found the legs and shoulders and tail of *Ichthyostega*. At last it had most of a body.

Jarvik, the expedition's expert on lobe-fins, was given the rocks and the task of reconstructing the animal. He could see that its legs, while short and squat, had the elbows, knees, ankles, wrists, and toes that qualified it as a tetrapod. He could find no sign of gills; its spine was sturdy, its hips and shoulders massive, its skull rigid. Yet Jarvik could make out a suture in the skull at the same place where, in a lobe-finned fish's skull, there was a hinge. Under the tetrapod palimpsest its ancestry could be seen. Its tail, on the other hand, was like a pennant reading FISH. Tetrapods have simple tails, a long series of tapering vertebrae encased in flesh (ours has dwindled to a mere sprout, the coccyx). A lobe-finned fish tail, which is the motor that the animal uses to move through water, is a much more elaborate affair. Each vertebra has two long rods, one on top and one below. Attached to each of these rods are more slender bones, called radials, and attached to the radials is a wide fan of fin rays: a completely different kind of bone called dermal bone that also makes up scales. This complex anatomy al-

Ichthyostega after a 1956 reconstruction by Erik Jarvik. (As we'll see later, paleontologists have since revised their image of this animal.)

lows the lobe-fin to set up waves in its tail either forward or backward, to let it dart through the water or suddenly brake. The bottom of *Ichthyostega*'s tail had a simplified tetrapod form, but the top still retained all the geegaws of a fish. It was, in a sense, still half in the water.

With Jarvik's work, paleontology had two milestones on the path that led our ancestors from the water to land: *Eusthenopteron*, a lobe-fin with faint fore-shadowings of our own bodies, and *Ichthyostega*, a basic tetrapod. It was not a pretty grandfather. Here was a three-foot-long beast, sprawling wide and dragging its tail along the ground, with eyes set on top of a head like a toilet bowl lid that could clap shut on prey, piercing it with its rows of fangs. Here indeed was a pleasant genealogy for humanity.

Chapter Two

LIMITLESS AIR, HO!

Around the time that Gunnar Säve-Söderbergh first raised up the bones of *Ichthyostega* from Greenland rock, other scientists had begun to wonder what combination of circumstances could have driven tetrapods onto land. The question was natural enough since its answer would tell us about our own heritage, but paleontologists at the time always managed to give it the flavor of Manifest Destiny. The vertebrates alive today whose ancestors never left the water are "but fishes," as one scientist denigrated them in 1916. "No matter to what degree they may have been specialized, they could not have risen nor can they ever rise to a higher plane. The emergence from the limiting waters to the limitless air was absolutely essential to further development and constituted one of the greatest crises in organic evolution."

Until recently paleontologists generally relied on a story of cruel droughts and other threats of death to explain how macroevolution got our ancestors through this crisis. It was brilliantly told by one of the century's leading paleontologists, Alfred Sherwood Romer. He was the son of an Associated Press reporter and grew up in White Plains, New York. A dog bit him when he was a boy, and he had to come to New York City to get a series of shots in his stomach for rabies. The shots took weeks, and he stayed with an old maiden aunt who didn't quite know what to do with a child. She brought him each day to the American Museum of Natural History and left him there, to make friends with the sail-backed dimetrodons and duckbilled hadrosaurs.

The friendship was a deep one. After serving in Europe during the First World War, Romer came back to the museum to study with William Gregory, a paleontologist who had championed *Eusthenopteron* as a close relative to the ancestor of tetrapods. Romer took up the work, studying how lobe fins could have been bent by evolution into limbs, as well as sectioning the skulls of lobe-finned fish to understand how the braincase was hinged. While teaching at Harvard, he spent his summers in Texas, digging out younger fossils of primitive amphibians and reptiles. His knowledge of vertebrate fossils was overarching, and in everything he did Romer carried on the traditions of both Gregory and of his journalist father—he saw evolution as a story, and his job was to write the missing chapters. He loved murder mysteries, and for his lectures, first at Chicago and then at Harvard, the auditoriums were always crammed. He had no help from appearances: in order not to be distracted by a decision over what to wear or the hunt for a matching sock, he always wore a black tie, black jacket, and black pants. And yet when this small man was done recounting some stage of evolution for his audience, they would be on their feet applauding. When he wrote an application for a grant, the judges would take it home for bedtime reading.

To explain the origin of tetrapods, Romer took as his inspiration a paper that had been published while he had been serving in Europe. According to a geologist named Joseph Barrell, many of the most telling clues were to be found not in fossils but in the rocks surrounding them. Barrell was an expert at identifying ancient landscapes from the color and texture of stone, and he was particularly gifted when it came to the Old Red Sandstone Formation of the Devonian. "Was it in the physical condition which determined the nature of the Old Red Sandstone," he asked, "or was it from the open sea that certain fishes grew to breathe the vivifying air, to crawl on solid land, and inaugurate from such humble beginnings that dynasty of terrestrial vertebrates which through all after ages was to lead in the march of evolution and rule over the living things of the Earth?"

Old Red was his answer. Its color had come while the stone was still soil, and iron compounds that it carried had rusted as they were exposed to the atmosphere. To create such rock, Barrell concluded, the continents must have been semi-arid and suffered droughts every year. The lobe-fins whose fossils paleontologists had found in these Devonian sandstones must have lived in lakes and rivers, and these waters must have annually evaporated to nothing. This kind of brutal environment could have pushed lobe-fins toward a tetrapod existence.

They would have been crammed into dwindling puddles during droughts, and the oxygen in the water would have run low or disappeared. "To breathe air would be an advantage, enabling the possessor of that power, although gasping for breath and half asphyxiated, to live where other fishes died in masses," he claimed.

Romer added some biological realism to Barrell's droughts. The tetrapods' closest living relatives are the lungfish, and they all live in fresh water; the fossil record in Romer's day suggested the same. Oxygen-poor ponds might be pressure enough for the evolution of lungs to complement gills. "But suppose the drought worsened and the water in a pond dried up completely?" Romer asked. "An ordinary fish would be literally stuck in the mud and would soon perish unless the rains soon returned. But a form in which there had been some trend for enlargement of fins toward the tetrapod condition—a lobe-finned fish—might be able to crawl up or down a river channel, find a pond with water still present, happily splash in, and resume his normal mode of life."

Legs were a way to go on living like a fish in an environment that became hostile to fishes. "Legs were, for the time being, simply an adaptation for bettering the animal's chances for surviving in his proper aqueous environment," he said. Later, when Jarvik slowly began to raise the curtain on *Ichthyostega,* Romer felt satisfied. It seemed sturdy enough to make its way across dry land to a pond, but it would be happier in the pond itself.

Romer told this story often because he considered it a good antidote to bad thinking. To some people the move from the sea to land seemed so drastic, so radical, that any change short of mystical didn't have a chance of working. Yet Romer had shown that adaptations to a life spent mostly in water might prepare an animal for life exclusively on land. Darwin had first noted this kind of unintended foresight, and its importance became even clearer in this century. Walter Bock, an ornithologist at Columbia University, looked at it with particular care in the 1960s, and his favorite example was the jaw of birds. When ocean birds like skimmers or pelicans hunt, they drop to the water's surface, open their mouths, and grab fish. At their speeds, dipping their jaw into the water creates a tremendous shock that they have to withstand with a second joint between the back of the jaw and skull. An ordinary bird on land has no such brace and would have its jaw torn off if it tried this stunt. Bock suggested a way that an ordinary jaw could evolve into its pelican form: Say a lineage of birds on land evolved stronger jaw muscles. A stronger muscle needs a wider base of attachment on bone, and so the back of the jaw would have to gradually develop a larger anchor for the muscle. In time the anchor would reach so far back that it would contact the skull and inadvertently form a brace. If these birds now began to experiment with jolting kinds of hunting—such as catching a fish—they were already prepared with a shock absorber that had evolved under another set of pressures.

For many years cases such as the jaw brace were called *preadaptations*. The word can give them an aura of conscious planning for the future that evolution can never have, and so in 1982 Stephen Jay Gould of Harvard and Elizabeth Vrba of Yale offered another term, *exaptation*. The word does a better job of capturing the notion of a structure crafted by evolution for one function and later hijacked for another. For example, the boneless ancestors of vertebrates may have needed to store phosphate in their skin to tide them over during lean seasons. The most reliable way to keep it was in a matrix of calcium, which happened to create a hard tissue—bone. A shell of bone also protects an animal from predators, and so natural selection began to craft this larder into armor. Later it even proved to work well as a scaffolding for muscles inside the body. So the production of bone—an adaptation for staying well-fed—became an exaptation as a skeleton. And in Romer's scenario, lobe-fins, with their primordial limb bones, were an exaptation for limbs that could walk on land. One can even think of animals as a whole set of exaptations. In Romer's case, tetrapods were fishes that were scurrying as fast as they could back to the water. Only much later did they use this water-to-water transportation to live on the land itself.

To get a glimpse of the kinds of places that Romer was talking about, these Devonian playrooms where baby steps were first taken, I made a trip to deep Pennsylvania. A paleontologist named Ted Daeschler had recently made a remarkable discovery: the shoulder of a tetrapod there in rocks 367 million years old, 4 million years older than *Ichthyostega*. He hoped to find the rest of it, and each month he tried to break away from his work at the Academy of Natural Sciences in Philadelphia for a couple of days. He didn't mind a naive field hand from Brooklyn helping out.

As I drove along the Susquehanna toward Williamsport, it was hard for me to imagine where anyone would even dream of finding fossils around there. From the stark pictures I had seen, Greenland seemed a sensible place to look, at least where it wasn't buried in ice, but in Pennsylvania life was too cozy. It would be futile to look in the hillside forests that spilled out onto the cornfields, to dig behind the grain elevators or the transmission shops, under the purple wildflowers or the Little League Museum or the sidewalk where a camp counselor pushed a shopping cart full of tennis balls. The only rock I could see was the gravel in driveways.

Daeschler met me at a parking lot along an airstrip outside of Williamsport. His face—solid, short-nosed, cheerful, bearded—reminded me of an otter's. We got into his Volvo loaded up with chisels and crowbars and rockhammers and blankets in the back, and drove to the West Branch of the Susquehanna, a narrow gorge full of slow trucks that was marked every few miles by fading towns: Queens Run, Farrandsville, Hyner, Renovo. Looking out the windows at the forests pressing in on both sides of the highway, I told Daeschler about how crazy searching for fossils here seemed.

"Soil is the bane of my existence," he said. "I go up to Hyner Point and I look out at the valley sometimes, and I can just imagine a big glacier coming down." He pushed his free hand forward toward the windshield. "It scrapes everything away down the river. You can see the site where I found the fossil, and it's just maybe a thousandth of one percent of the whole formation. You'd get a lot of till at the bottom of the valley, but all the walls would be scraped nice and clean." Rather than wait for the next ice age, Daeschler depends on the Pennsylvania Department of Transportation. For thirty years they have been blasting through hills and mountains to build wider and straighter roads for the trucks and station wagons that flow through the state. You can't drive too far through Pennsylvania before passing the giant blinking arrows squeezing you over a lane, the bored-looking girls in hard hats flicking their orange flags.

Daeschler first saw these rocks in the early eighties when he was majoring in geology at a college not far away. They are part of what's known as the Catskill Formation, which stretches from New York down to Virginia. These were some of Barrell's original Devonian parched landscapes. At the time, Europe was pinning Greenland against the east coast of North America, pushing up raw new ranges of the Appalachians as volcanoes blasted lava across New Hampshire. On a map the assembled landmasses look like an equatorial blob, with the east coast of the United States extending from it as a long, mountainous cape. The range cast sediments down its western slopes, across a broad alluvial plain to the shallow Catskill Sea, to underwater Ohio.

After getting a master's degree and spending time digging up fossils of mammals and dinosaurs, Daeschler got a job back in Pennsylvania. As the collections manager at the Academy of Natural Sciences, he kept the fossils there in order, and to stay up to date, he took the bus over the Schuylkill River to crash paleo seminars at the University of Pennsylvania. After a few years he decided it was time to get his Ph.D. He needed a field project for his dissertation, but it couldn't be anywhere too exotic because he and his wife had just had a baby and he was still holding down his Academy job. There was the Catskill Formation, which had yielded a few interesting fossils. One of the first fossils of a lobe-finned fish had been found by a railroad engineer near the New York border in the 1840s, back when the cuts were blasted for train tracks rather than freeways. Romer had always thought there should be tetrapods in the Catskill, and one of his students, Keith Thomson, had worked one summer in the 1960s on some of the cuts on the road between Hyner and Renovo. There Thomson had managed to find some bones of a nightmarish Devonian lobe-fin he named *Hyneria*, with fangs the size of thumbs and a body longer than a Cadillac. In the seventies the road was widened another few yards and Thomson, now a professor at Yale, had sent a student back to look at the fresh rock. After a whole summer he found only scraps of lobe-fins. Thomson later came to the Academy to serve as its president, and when Daeschler asked about the Catskill, he didn't think there was much to be found.

Daeschler decided to go anyway in 1993. He surveyed the red mudstones along the roads and slept in a tent to keep the gamble low. We passed the road to the campgrounds where he had stayed, and it started to rain. "On the first day it rained and rained like today. Miserable," he said as he squinted past the windshield wipers. "There was nothing in the red bed. The next time I found really crappy stuff."

To wait out the rain, we drove past Daeschler's site, across Young Woman Bridge, and on to Renovo. On the east edge of town, where the Allegheny Plateau dropped down to the flood plain of the West Branch, Victorian brick temples of industry still stood, where for almost a century railroad cars were welded together. A line of orphaned cabooses sat on the rails; a local businessman had an idea to sell them as vacation homes. The sidewalks were empty, the curbs unparked. Daeschler and I ran through the rain into a jewelry shop owned by a man named Norm Delaney, who sometimes digs with Daeschler. Fossil lobe-fin teeth are on display in his window along with the emerald rings.

When the sky brightened, we drove back to the site. Another digger named Doug Rowe met us there. Rowe was an engineer at the railroad car factory until it shut down, and now he made do as a night manager at a hotel in Renovo. He is the kind of man who will tell you exactly which kinds of material a porcupine likes to eat—T111 plywood, aluminum, tires, plastic steering wheels—but says it in such a quiet way that you know he's right. Walking one day on the north end of Renovo he found a rock with the image of a leaf pressed into it. A few days later, he saw Daeschler high on a road cut called Red Hill, and soon he was spending more time on the rocks than Daeschler himself. Now he can tell you about the nearly invisible scale of a fossil ray-finned fish with almost as much certainty as about porcupines.

Daeschler and Rowe pulled out the tools. The Red Hill rock face shot up a hundred feet bare before its slope softened into trees. The top half was Carboniferous sandstone devoid of fossils, and underneath were layers of Devonian mudstones, reddish and bulging out of the side of the hill like loaves of bread. This was where Daeschler had first come in the rain; he had driven to other roadsides in the weeks that followed, and finding nothing he came back for another look. He was beginning to think of backup plans to salvage his Ph.D., maybe turning himself into a paleoecologist and using whatever fossil scraps he might find to guess at the Devonian ecosystem here. As he stood along the road with Rowe and me, he picked up a rock from the asphalt and winged it up at the road cut, hitting a hammered-out scoop in a section of greenish rock. This is how he points.

In that greenish rock, he explained, he had luck. He came across a few fossil scales the size and shape of potato chips, which belonged to a lobe-fin. Now he had at least a little hope of finding fossils worthy of a dissertation. He dug around the scales and coated them in plaster so that they wouldn't crumble on the way back to Philadelphia. As the plaster dried he killed time by inspecting other parts of the layer—"just prospecting, poking around," he said. "And not more than three meters away, in the sort of place where you have to hang with the tips of your boots and your fingertips, I saw a flat bone in cross section." Carving himself a small ledge to stand on, he began to work off the weathered rock from the top of the fossil with an awl. It was a teardrop-shaped blade of a bone that Daeschler recognized as a left shoulder. Yet it wasn't like the shoulder of any lobe-fin he could remember, and the following morning he found a second left shoulder three inches from where he had found the first. He wrapped the bones in tissue paper and masking tape and got in his car to drive three hours back to Philadelphia. Something was eating at him: if these bones didn't belong to a lobe-fin, they might be the shoulders of tetrapods.

With the help of Keith Thomson that night and paleontologists at Penn the next morning, he studied the bones. They opened up Erik Jarvik's vertebrate textbook and compared the shoulder to the picture of *Ichthyostega*. When a fish breathes through its gills, it opens a pair of flaps to let the water flow out. When the flaps close again, they clap tightly against the shoulders along a flanged groove. But Daeschler's fossil shoulder—shaped like a three-inch trowel—was grooveless. It also bore a raised slot on the bone like the socket in *Ichthyostega*'s shoulders. This new creature was rugged: its shoulder had broad scoops and buttresses where muscles would have attached, muscles that would have lashed the shoulder to the spine and let it haul its arms forward. Daeschler couldn't quite believe it: in this miserable, worked-over rock, he had found a tetrapod, the only Devonian tetrapod from North America outside of Greenland, and among the earliest in the entire world.

Success made life a little easier. When Daeschler published his new tetrapod (which he named *Hynerpeton:* creeping animal of Hyner), he was able to get the funds for a backhoe complete with a giant jackhammer to putter out to Red Hill in 1995. Having found two left shoulder bones—obviously from two individuals—so close to each other, he thought he might have found the edge of a bone bed littered with tetrapod skeletons. The jackhammer pried away tons of the overburden above the layer to give him quick access to the green rock. For days he searched it, but he couldn't find another bone. Chances are that the crews who had made the road cut dynamited the rest of the *Hynerpeton* skeletons into pebbles. "That's paleo," Daeschler said, shrugging, and we walked back south along the road.

But the next year Daeschler, Rowe, and Delaney found another lens of rock a short walk down the highway that was so crammed with fossils they named it Big Fish Alley. This was our destination as we portered Daeschler's tools up the cut, working our shoes into the red chips. On a ledge that Rowe and Delaney had carved, we broke off a block of mudstone the size and shape of a couch cushion, first banging chisels down into the rock and then working the ramrod and prybars underneath to move it away from the wall. It collapsed into a few steak-sized chunks, and we each took one.

"Find the place where the rock wants to break and encourage it," Daeschler told me. I laid it on its side and worked a pinky-sized chisel between two layers of sediment. There was no noise except for our tapping and the metal guard rails along the highway popping in the late summer heat. With a few taps, my rock split open with a crisp rustle. I laid the two pieces on the ledge as if I had just cut open the pages of a book of illustrations. On these pages I saw stems and tufty leaves belonging to one of the first trees on earth. Its name was *Archaeopteris*, and along rivers it grew sixty feet high, a tall conifer shaft with these

Archaeopteris

fronds hanging from its branches, overshadowing the life below. I opened more pages and found the fruiting bodies of *Archaeopteris* and leaves so well preserved that I could pull up the veins. My fingers turned black with their carbon, and I exposed bark still puckered around the pores through which the tree breathed. Scattered among the *Archaeopteris* were minnow-sized fishes, iridescent lungfish teeth that crumbled at the touch, scales of armored placoderm fish. Terrestrial scorpions and relatives of ticks have turned up in Big Fish Alley, and Rowe has even found the jaw of a tetrapod here. It can only mean good news, because it is either *Hynerpeton*'s jaw or the jaw of the second Devonian tetrapod species from North American. In the few hours I spent on Red Hill, I didn't find any new tetrapods myself, but at least I touched the trees that cast shadows on their backs.

Daeschler has brought paleontologists of all stripes up to Red Hill: palynologists studying fossil pollen, paleopedologists who study fossil soil, experts in fossil vertebrates and invertebrates. They all plead for him to save fossils for them, with the result that the operation at Big Fish Alley is now more like triage than a fossil dig. It's now clear that Daeschler has exposed the best picture of the ecosystems in which the first tetrapods lived. Giant rivers as big as the Ganges as well as smaller streams meandered through this place 367 million years ago. In some places they formed oxbows where *Archaeopteris* and other plants dropped their leaves and branches, and where animals left their remains and were covered in muck. Here lobe-fins were royalty, scarfing up the smaller fish. Ferns as tall as a man grew along the water; club mosses sent their roots into the ground. Monsoons swept across Pennsylvania, and the rivers and streams would swell and haul away tons of muck and wood to the ocean close by. The forests would flood and stay submerged for months, *Archaeopteris* surviving on high banks and ridges, and then the waters would subside into their banks again. "Pennsylvania in the Late Devonian was not a wasteland," Daeschler said. "It was as swampy as Louisiana."

How could the drought-wrecked horizons of the Devonian turn out to be a mirage? One paleontologist I spoke to put it this way: Barrell saw two and two and made twenty-two. Red rock can be formed in a parched landscape, but since Barrell's time geologists have come to recognize that it can also be created in many other ways. If soil rich in ferric oxide is carried down a tropical river through a lush landscape, it can color the rock formed where it is deposited if it simply has a little time on the surface to rust first. There were certainly dry stretches of Devonian earth, but much of it—particularly where the first tetrapod fossils are found—was lush and moist.

The other part of the old scenario—how tetrapods evolved in freshwater ponds and rivers—also crumbled, thanks in large part to Keith Thomson, Ted Daeschler's advisor. By the time Thomson began to make lobe-fins his specialty in the 1960s, scientists had learned that lungfishes were not the only living lobe-fins after all. In 1938 fishermen from South Africa had pulled a four-foot-long creature out of the Indian Ocean, and scientists from the East London Museum recognized from its fleshy fins and tasseled tail that it was a form known as a coelacanth. Coelacanths, like lungfishes and *Eusthenopteron*, are lobe-fins, their roots reaching back to the Devonian. Until 1938, though, paleontologists thought that they had vanished 65 million years ago with the dinosaurs. The dead, gutted creature that came to the museum told the scientists that coelacanths must be still alive in the deep ocean, but it wasn't until 1952 that the next specimen surfaced. More recently biologists have descended four hundred feet in submersibles to watch them swim slowly through dark water, snatching cuttlefish and other prey, as their fins move left, right, left, right, in a distant deep-water echo of a tetrapod trot.

To Thomson, it seemed that the coelacanth was radically out of place. "The living lungfishes live in freshwater," he says, "the living amphibians mostly live in freshwater—there are a couple of marine toads and whatnot, but otherwise they're freshwater—and so I followed everybody else's logic, that if the living amphibians are descendants of the lobe-finned fishes, then the lobe-finned fishes were freshwater too." If that were so, how could a coelacanth live not just in salt water, but far below the surface of the ocean? Thomson had found his fair share of lobe-fin fossils near Hyner in Pennsylvania, and they hadn't lived in some freshwater inland lake, but in a river delta feeding into the ocean, a delta that was probably briny. He reviewed the fossils that had been collected over the decades, and while many species lived in freshwater, the oldest lobe-fins as a rule came from marine rocks. Thomson began to contemplate a different birthplace for tetrapods, one that has since been supported by much research over the last thirty years. They originated not in drying ponds (which probably weren't drying anyway), but in coastal lagoons.

At the time, these wetlands were an ecological frontier. Before about 480 million years ago, the continents were lifeless save for thin crusts and varnishes of bacteria. Green algae growing near the coasts began to creep into inlets and estuaries, where rivers carried fertilizing loads of weathered minerals and dead microbes. The algae moved into shallower waters and evolved a thick skin that protected against longer and longer exposures to the open air, until they no longer had to be submerged. They became land plants. Soaked by sunlight, they evolved into low mossy mats and then ground-hugging weeds. At first they were limited by simple plumbing; their stalks weren't strong enough to grow high, nor could they pump nutrients against gravity. New species evolved, slowly at first, developing roots and associations with underground fungi to extract food and water from the ground, and hard tissue to hold botanical weight high. When the plants died their corpses enriched the estuaries and inlets with nitrogen, phosphorus, and other scarce elements, making it easier for succeeding generations to grow. Invertebrates emerged from the water, beginning with millipedes 450 million years ago, and then followed by insects, spiders, scorpions, worms, and mites; they began to root through the thickening soil and nibble the stems. They barely checked the spread of plants, though: ferns, club mosses, and horsetails took root, wood wrapped around plants and leaves unfurled, levels of carbon dioxide plummeted in the atmosphere as it was turned to tissue, and by 370 million years ago, in the mid-Devonian, full forests were growing and spreading inland.

These young coastal wetlands may have been the original home of lobe-fins. Not shaped for long, deep-ocean swims like sharks and tuna, lobe-fins were better suited for sudden bursts through sludgy water. One of the risks of life on the coasts was the sudden drop in the oxygen caused by blooms of bacteria feasting on organic matter in shallow water and sucking up the gas for themselves. But with lungs the lobe-fins could get their oxygen elsewhere. The fossil record suggests that lobe-fins thrived in these places. Evolving into enormous forms, they feasted on the galaxies of small fishes that in turn fed on the invertebrates that lived in the muck. Their limblike fins even let them move along the bottom of the lagoons, and in shallow water they might have pushed up with their front fins to breathe. Barrell and Romer thought that tetrapods had to have emerged in a brutal world where only change could keep death at bay. But to Thomson, like other paleontologists of his generation, macroevolution didn't have to be so harsh.

In carrying the lobe-fins back to the ocean, Thomson may have also explained the origin of one of the most important parts of life on land: the matter of urine. Whether we live in Sumatra or the Sonora, all tetrapods survive in the desert of air. We are made mostly of water, which desperately tries to get out of our bodies, to do its little part to bring a moist equilibrium to inside and outside, and we do what we can to keep it trapped within skin and scales. Mammals and birds—animals with high metabolic fires and rapid breathing—pack their nostrils with scrolls of tissue that can grab water before an exhalation carries it away.

All this water-trapping runs head-on into conflict with some of our most common metabolic reactions. We eat food full of nitrogen—more nitrogen, in fact, than we really need—which can hook up with hydrogen atoms, turning into poisonous ammonia. Because ammonia dissolves so well into water, we would have to urinate gallons to flush it out through the bladder. Our physiology makes a smarter choice: in our livers we join together ammonia and carbon dioxide with the help of enzymes, creating a much safer compound called urea. (This same chemical cycle also generates an amino acid essential for building many proteins—which might hint at how the production of urea evolved in the first place.) Tetrapods can let the urea build up in the kidneys without fear of poison and then get rid of it in a concentrated flow of urine. Some animals, such as those that live in dry habitats, have to be even more careful with their water and turn the urea into a powder before they excrete it.

Fishes use the same basic metabolism as we tetrapods, and they churn up just as much extra nitrogen. In a lake or river where dehydration is not a worry, ray-finned fish simply flush the ammonia out of their gills and drink more water. (The African lungfish uses this strategy when it is submerged in a river, but after it makes a burrow during a drought it switches to making urea to conserve water.) In the sea it's not so easy to dump ammonia because the water is loaded with salts—not just the sodium and chloride of table salt but other charged atoms such as magnesium and sulfur. Vertebrates use some of these ions in their nervous system to propagate signals. Thus a fish that flushes lots of fluid out of its gills and takes in seawater to replace it may flood its neurons with confusion. Ray-finned fishes that live in the ocean pump ammonia through their gills and fight against the encroaching salt they drink by constantly excreting it through special glands.

Sharks, on the other hand, survive in the ocean by doing something else altogether: like us, they turn ammonia into urea. Urea makes blood salty, but unlike the charged atoms in seawater, it doesn't interfere with the nervous system. By keeping their bodies safely salty with urea, sharks aren't swamped by nerve-poisoning charges. And in 1966, when Thomson and his fellow biologists at Yale analyzed the chemistry of fresh coelacanth blood for the first time, they discovered that it was also full of urea, as much as in a shark.

Before Thomson's work, our own urea was a puzzle: if tetrapods evolved from freshwater lobe-fins, they must have evolved it from scratch. But when Thomson considered all the evidence at his disposal, the simplest explanation was that the earliest vertebrates—the ancestors of sharks, lobe-fins, and ray-finned fish—had invented urea to survive in the ocean. Sharks that stayed in the ocean simply went on producing it, while lobe-fins, which evolved in briny coastal lagoons, still depended on urea to avoid being overwhelmed by salt. The coelacanth, later moving out to deeper waters, held on to the cycle, while African lungfish shut it down for the most part as they lived in freshwater, although they could still resort to it. By contrast, ray-finned fish moved to fresh water and lost their ability to produce urea; later, when some of their descendants came out to the ocean, they couldn't recover it. They had to invent glands instead to fight salt.

Our own origins make much more sense in this arrangement. When our lineage of lobe-fins left salt water for dry land, ammonia poisoning and water loss still plagued them. Fortunately they were already prepared with the exaptation of urea. Thomson helped to dismantle much of his old teacher's vision of the origin of tetrapods, including the supposed exaptation of legs that evolved to let them walk back to water. But with his work on urea, he found an exaptation to take the place of the one he had helped spoil. Water and air are, at least in one sense, actually not all that different. In going from the sea to land, we have traveled from one desert to another.

When scientists like Romer and Thomson roughed out their hypotheses about the origin of tetrapods, they had to work around a stupendous gap in the fossils. The trail of tetrapod evolution led up to the lobe-fins like *Eusthenopteron* and then leaped forward to *Ichthyostega* with full-blown legs and other traits of tetrapods. As the century aged, *Ichthyostega* still stood alone. A few fossils of Devonian tetrapods emerged around the world—an interesting jaw from Australia, a few scraps of something from Brazil. But by 1982, the fiftieth anniversary of the discovery of *Ichthyostega*, the silence of the rocks had become maddening.

During this time, paleontologists who wanted to study this transition had to content themselves with working on either side of this gap. One of them was an Englishwoman named Jenny Clack. After more than fifteen years of false starts and frustrating dead ends, she has begun to fill the fish-to-tetrapod gap by finding some of the most important fossils in all of paleontology. And yet when you meet her, as I have at scientific meetings, this small, salt-and-pepper-haired woman is soft-spoken and reluctant to regale. As a girl growing up in the fifties, Jenny (then known as Jenny Agnew) was drawn to fossils but not the ones that typically attracted kids. "I'd always been interested in the earlier parts of the history of animals more than the dinosaur end of things," she says. Her favorite fossils were unlike anything alive today, such as armored, jawless fish or the ribbed trilobites that foreshadowed horseshoe crabs and scorpions.

In the sixties she studied as an undergraduate at the University of Newcastle-upon-Tyne in northern England with the paleobiologist Alec Panchen. Panchen and a coterie of graduate students had studied lobe-fins as well as some of the tetrapods that appeared post-*Ichthyostega*, after a gap of some 30 million years. In that gap a macroevolutionary bomb must have gone off, because over a dozen different kinds of tetrapods have been found living in those Carboniferous swamps. The Newcastle paleontologists studied amphibians with heads like boomerangs, others with bodies like snakes 200 million years before the first snake was born, others still with bodies flattened as if bank safes had fallen on them. Yet the skeletons themselves are some of the rarest of all fossils. As Jenny

Agnew finished up her degree, she hoped she could work on these creatures as a graduate student of Panchen's. "I approached him to see if there were any opportunities going. He said not."

Eight years passed. She learned how to make a living in museums and went to work in Birmingham. She built displays of insects and birds, brought schoolchildren through the exhibits on holidays. She met Robert Clack, a man who was even more quiet than she, who programmed computers for a living and spent a fair amount of his free time looking for fossils in quarries, and they married. Sometimes they would ride their motorcycle through Scotland, stopping along the road to inspect the Devonian rock. For no particular reason, her boss noticed in 1979 that Clack was getting tired of her work and suggested that she find some kind of special project to work on for the museum. Clack went back to Panchen to see what opportunities there might be.

This time Panchen had an idea. There was a tetrapod by the name of *Pholiderpeton*, and it came out of a 330-million-year-old bed of coal in Yorkshire in 1870. A crocodile-looking creature, it was somewhere close to the origin of amniotes—the animals that today include birds, mammals, and reptiles. For a century it had sat in a Yorkshire museum, still lodged in blocks of coal, which were sunk in a plaster mount. Panchen had tried to borrow it and had been turned down. Luckily for Clack, the museum was going through upheavals and when she asked, the staff was actually happy for her to take away this fossil for good. She brought it to Newcastle, where she thought she would do nothing more than free it from the coal, describe what little was there for a few weeks, and bring it to the Birmingham museum when she returned to her regular work. But when she took off the plaster from the coal and turned it upside down, she discovered the animal's braincase. Those hidden bones switched Clack's life onto a new set of tracks. Only smashed bits of these animals had ever been found before, and not long after she showed Panchen what she had found, she was transformed at last into his Ph.D. student.

Clack spent three years at Newcastle reconstructing *Pholiderpeton*, paying special attention to its ears, or rather to the blobs of bones that would later evolve into ears as we know them. By 1981 she had landed a job as an assistant curator at Cambridge University's Museum of Zoology and spent a few more years finishing up her work on *Pholiderpeton*. But as her work came to a close, she began to wonder what she could do next. She was now an expert on describing early tetrapod skeletons, but there was no orphaned fossil waiting for her to claim. There were rumors of a newly found Devonian tetrapod in the Soviet Union called *Tulerpeton,* but she had no hope of going behind the Iron Curtain to inspect it. By the mid-1980s Panchen and other paleontologists were beginning to make sense of their Carboniferous tetrapods, sorting out which were most closely related to each other and to today's living species. But they were still stymied over how the tetrapod body plan had evolved in earlier forms.

If Clack could have looked at *Ichthyostega* she might have found a few answers, but there she was stuck. Ever since Gunnar Säve-Söderbergh died, the task of describing *Ichthyostega* belonged to Erik Jarvik, and he still hadn't finished. Some paleontologists, perhaps because they are so accustomed to talking about the passage of millions of years as they rest their hands on a layer of rock, are often slow to publish their research. And to Jarvik, a scientist who looked at life from a fish's point of view, *Ichthyostega* was merely a lobe-fin with fingered lobes. He studied dozens of fossil species, establishing new divisions for the lobe-fins, and he spent twenty-five years grinding the skull of *Eusthenopteron,* working down a millimeter at a time, drawing and photographing each fresh surface, using the pictures to build wax models of the creature's brain and gills. From time to time he published sketches and photographs of a few bones of *Ichthyostega* in the tomes he wrote, but his final monograph came out only in 1996.

In the early 1980s Jarvik was still over ten years away from his last word on *Ichthyostega,* and paleontology's unwritten rules demanded that no one publish any new information about it until he did, no matter how long it took. "It was a delicate situation at the time," says Clack. "He had been working on *Ichthyostega* since the thirties and forties, and he hadn't really completed the work. He had come out with one or two publications—one big one which described the tail and very little else. He'd been sitting on this stuff. It's not that he wouldn't let anyone else look at it, but they weren't at liberty to publish anything on it. One respects other people's territories, and that was regarded as his."

What to do? A voice began to whisper a name to her: Greenland. The voice belonged to her husband. When he and Jenny would ride their motorcycle along the coast of Scotland, they half hoped they'd discover tetrapod fossils, but they found only scraps of fish. He would imagine exploring the old territory where Säve-Söderbergh and Jarvik had walked, the only place on earth where full Devonian tetrapod skeletons had been found. "To her it's material to work on, but to me, since I'm outside the academic world, it's something to fantasize about," says Robert. Jenny wondered if perhaps he didn't quite understand how paleontology worked. Trips to Greenland are not like trotting off to the local quarry; they are expeditions, with helicopters and rations and two-way radios. Jenny, a museum creature, knew she was no mountaineer like the Scandinavian scientists who had gone there. And then there was rationale to think about: no one would give her the money to go to the sites in Greenland that Swedish scientists had picked clean for decades, nor would they think much of Clack wandering aimlessly, fjord to fjord, in the hopes of finding something.

Still she began to toy with the idea, to pretend at least that it might not be naive. "It was Robert's prompting that made me follow up the leads, such as they were," says Clack. After the Swedes stopped going to Greenland, a team of Cambridge geologists led by Peter Friend continued the work through the 1960s and 1970s. They weren't interested in fossils as anything more than ways to mark the layers of rock they were handling, but Clack wondered if they might have come across some chip of *Ichthyostega* from a site where the Swedes hadn't been. It was no great expedition for her to walk to the Sedgwick Museum where Friend worked. When Clack told him about her curiosity, he hauled out a pile of literature and wished her luck.

She worked her way down through the stack. She read journal papers, government reports, field notes. She read about how in the 1970s one of Friend's students named John Nicholson climbed up the southeast slopes of Stensiö Bjerg, a mountain north of where Jarvik had worked. The reports detailed the Devonian stratigraphy and then listed the fossils he had found. Wedged among them—the antiarchs, the lungfish, the other lobe-fins—was the casual note, "we found Ichthyostegid bones." Clack snapped to attention. She looked through Nicholson's field notes. "Tetrapods," he wrote, "are common here."

Clack walked quickly back to the Sedgwick Museum and into Friend's office. As calmly as she could manage, she asked if he still had the skulls. Nicholson had gone to work for an oil company in Scotland ten years earlier, but he had left his rocks behind, stored in cabinets in the basement. Friend led Clack downstairs and tracked down the numbers on the rocks. He pulled open a drawer, and sitting there were blocks of sandstone in which she could see skulls. They had been pretty badly worked over by wind and rain, the fine ornamentation blasted clean off, the spongy inner bone eroded until it looked like caramel. Yet she could see that these skulls, which were smaller than those of *Ichthyostega*, had a pair of prongs projecting from their back. They made Clack start: *Ichthyostega* doesn't have prongs.

She knew that she was looking at a fossil ghost. Fifty-three years earlier, on Erik Jarvik's first summer with the Greenland expeditions, he had traveled with

Säve-Söderbergh along Gauss Halvo's southern shore, as Säve-Söderbergh named mountains after men and valleys after women. They found the bones of lobe-fins and *Ichthyostega* at most sites there, but on Wiman Bjerg they found half of a small tetrapod skull with prongs. They knew it was clearly a different genus from *Ichthyostega*, perhaps even a different family, but nothing more. Only in 1952 did Jarvik publish a terse report on it, calling it *Acanthostega* ("horn plate"). No one had ever found another bone. "I still remember Säve-Söderbergh sitting for hours in our tent, twisting and turning this specimen with a puzzled face," Jarvik wrote in 1996.

And here were not just more bits of skull, but three nearly complete ones, with more left behind on a mountain that Säve-Söderbergh and Jarvik had never explored. She got wind of a Danish geological expedition to the region to look for oil, and with her new skulls, she managed to talk her way into coming along. Events flowed as fast as they do in dreams: a paleontologist named Svend Bendix-Almgreen from the Geological Museum of Denmark organized an expedition to find *Acanthostega,* and in 1987 she found herself aboard a helicopter flying alongside bare mountain islands, with Robert, Bendix-Almgreen, and her graduate student Per Ahlberg. The helicopter set them down fifty miles from the base camp on the bare slopes of Stensiö Bjerg.

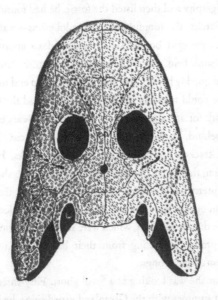

The skull of *Acanthostega,* viewed from above

They set up tents on a broad ledge 270 feet above the ocean. The never-setting sun circled them, sinking and rising like Andrée's failing balloon. They began to search for Nicholson's site, but his logs showed him going straight up the side of the mountain, which required immediately scaling a high sheer cliff. "We couldn't see how on earth he'd done that," says Clack. They chose an indirect route, swinging up a gentler climb over a bluff that brought them above the site, at a point where they could make their way back down. On the first climb, as the British beginners tried to walk on the solid rock and avoid the scree of fallen rubble, the Danes quickly left them behind. "You realize that the scree slope is the safe place to walk on because the rest of the mountain is called rotten rock," says Clack. "If you try to get a foothold on it, you just break it away and tumble down. Whereas the scree you can just paddle through."

On the rotten rock it took them four hours to climb the mountain; on scree it took them two. Each time they tried to find Nicholson's site but encountered nothing but blank stone. After days of searching they wondered if they were on the wrong mountain altogether, but before they gave up they decided to consider the possibility that Nicholson had used a broken altimeter. They hiked all the way down to the overhang that Nicholson claimed he had scaled and then zigzagged back up slowly. As they walked they scanned the rock, occasionally tricked by calcite and weathered bird guano. Then Jenny noticed a piece of the scree marked with real bone. It was part of an *Acanthostega* skull. More fragments lay on the slope, forming a trail leading uphill, and they followed the rock until they found where it was coming from.

Just as Nicholson had written, tetrapods were common here. The crew filled their backpacks with rocks that held *Acanthostega* skulls, legs, spines, and tails. At the camp they laid them out on newspapers, tagged them with sticking plaster and labels, and set them in the wooden boxes in which their food had come. After two weeks they had scoured the site clean, and in the time they had left in Greenland they explored the fjords nearby. There they found lobe-fins of all sorts and even a hind leg of *Ichthyostega* better preserved than Jarvik's own. By the time they were finished, they had collected half a ton of rock that was crated, helicoptered, shipped, and trucked back to Cambridge.

As is almost always the case, finding bones in the wilderness was only preamble. Clack's fossils were lodged in viciously hard rock, accessible only with diamond-wire saws and drills and air-powered mallets. Clack had personnel problems as well: Per Ahlberg was leaving her to take a postdoctoral position at Oxford, where he would busy himself with the lobe-finned fish they had found in Greenland. If she hoped to get the news of *Acanthostega* out before her retirement, she needed help. She hired a preparator named Sarah Finney, and she brought in a paleontologist named Michael Coates, whom she had met at Newcastle-upon-Tyne. He had become an expert on ancient vertebrates there, having studied fossils of primitive ray-finned fishes that had been found in a stream bank at a Glasgow playground. After finishing his Ph.D. Coates discovered just how dry a field paleontology can be. "Shall we say, I spent a few years at home in the human development industry," he says. He was close to packing it in as a paleontologist when Clack asked him to join her. The question was almost absurd. "Devonian tetrapods are rare as gold," Coates says, "and you know that anything you do with them is going to shake the whole tetrapod tree, because you're working down there at its base."

Clack liked the fact that Coates was a fish man, someone who considered tetrapods as only one variation on aquatic vertebrates. There was no telling how fishlike *Acanthostega* might be, and he knew all the esoteric bits of fin and skull on something like a lungfish that Clack would miss. Coates was just as relieved that she would handle some of the finer points of being a tetrapod. "She could say, 'Oh Christ, gill skeletons, those are all those horrible bits, and thank God tetrapods don't have them.' I can say quite similarly from my point of view, 'Stapes, oh God, thank goodness I don't have to deal with those—all those extra doodads and holes and getting seriously worried about where the seventh cranial nerve goes. Who gives one?'"

The two of them—Clack & Coates, like a Dickensian firm—set up shop in her Cambridge lab. Even before the bones had been freed from the rock, they divided up *Acanthostega:* to Clack went the skull, and to Coates everything else. It might seem like Clack was getting short-changed, but in paleontologists' currency she was getting at least an even split. From the neck down, a vertebrate's body is big-boned and repetitive—see one lumbar vertebra, you've seen them all. For those hoping to understand how an animal lived, the head is where thinking, hearing, seeing, smelling, and eating took place. The skull is also a Talmud of esoteric details, of little bumps, holes, flares, and curves, that can bind together families of creatures that an entire leg often cannot. And while the skulls of *Acanthostega* were clear on the surface of the rocks, the paleontologists couldn't tell how much of the rest of the skeleton was buried inside them.

Clack and Coates began to work on the bones in 1989. "I remember toddling on down to Cambridge in the beginning of July," says Coates, "and at that point my family was still up north so I didn't have to get home to put people to bed, and while I was having this second bachelordom I could piss around in the lab at all hours." He had to choose which bones to start on. There were five skulls exposed in their rocks. In one block were Spot, Patch, and Fido, named for the way they looked like puppies in a basket. ("Spot is named as a perverse joke at the expense of *Spot the Puppy* children's books, which are very popular in the U.K.," Coates kindly explains. "I fancied Spot's mum going to the kennel and finding the gruesome carcass of *Acanthostega*. When you've read five hundred Spot stories, these things happen.") There had been a movement to name another skull Bambi because of its eyes, but Coates vetoed it. Instead, he christened it Grace, after Grace Jones, for the way a layer of rock that was glued to the top of the skull looked like her flat-top hair. The fifth was Boris, which was the closest Clack could come to Boreal. Boris looked the most promising. "We knew that this particular one was what you'd hope for," says Coates. "Jenny had looked at the skull of it, and if you put all the bits of that rock together and you looked at it with the late afternoon sun coming through the lab window, you could see all these little lumps that looked very hopeful, that looked like they were the vertebral column extending just under the surface of the matrix."

One hope that Clack and Coates had for the fossils was for digits. This transformation from a lobe fin to a hand, complete with wrist and fingers, was one of the most important changes on the path to the tetrapod body. Five is the standard number of digits for tetrapods, a rule broken only by various species of amphibians, reptiles, and mammals that have lost one or more. Coates had to wonder, why of all numbers did a lobe-fin lineage choose five? In the block that held Boris's skull he and Clack saw a humerus, the upper bone of its arm. It was a flat, sturdy slip of a bone with the sort of smooth surfaces on which muscles could attach—something like the upper arms of other early tetrapods. On the adjacent piece they could make out two bones in cross section, which had to be the ulna and radius of the forearm.

"So the first thing I did was go on a couple of practice runs, just drilling around," says Coates. "I got the humerus cleaned up, and then I went down and found the radius and the ulna going to the end of the block. And the ulna was complete and very stubby and short, and the radius was long." By these bones alone, he knew that *Acanthostega* was clearly much different from *Ichthyostega* or from any other tetrapod. Hold your arm out, and your radius and ulna stretch away an equal distance. *Ichthyostega*'s is no different—an equality that's essential for walking on all fours on land. *Acanthostega*'s arm, however, was badly skewed for a tetrapod—the radius was twice as long as the ulna—but perfectly in keeping with lobe-fins. "The humerus looked conventional for an early tetrapod, but the radius and ulna, they could have almost come off a fish." Already Coates and Clack knew that *Acanthostega* would grab the attention of other paleontologists. "We looked at that and thought: paper—instant paper."

Still there was the question of hands. Coates chased the radius from the elbow down into the rock until he met up with a finger. It was a healthy tetrapod digit, which Coates also found odd on such a fishlike arm, but he didn't have much time to ponder. "I went through the digits like counting the numbers on a clock face, just working my way through." He could only dream that this fossil had all five fingers on its hand. The Swedes hadn't managed to find very good fossils of *Ichthyostega*'s limbs, and so Jarvik had only assumed that they had five. "It was riveting. I kept going, eighteen hours a day, just sweating it, thinking, 'Christ, it's coming on another digit.' It was hot oil and all this crap flying off the dental mallet, forging ahead desperate to try and find how many fingers were going to be at the end of this thing."

That night he opened his diary. "Worked late. Completed prep of *Acantho* forelimb. Eight digits." Coates had been wrong—the tetrapod tree wasn't shaking; it was beginning to flail.

Chapter Three

HOW TO MAKE A HAND

After hiding for 363 million years on a Greenland mountain-side, the eight fingers of *Acanthostega* managed to emerge with perfect timing. Only three years before Mike Coates scraped away their obscuring sandstone, other scientists had pushed aside the dominant ideas about how limbs evolved. They drew their inspiration not so much from paleontology, though, as from distant sciences—embryology, for example, and even mathematics. Their work cleared an intellectual space in which Coates and others could accomplish something that had never been done before in the study of macroevolution: to use fossils to reconstruct how genes turned lobe fins into tetrapod legs. If *Acanthostega*'s strange hand had appeared even a decade earlier, however, no one would have been ready to understand it.

Richard Owen would certainly have been scandalized. He thought that our five-fingered hands, like the rest of our skeleton, were based on a vertebrate Archetype—a skeleton that consisted of a string of vertebrae with various knobs and prongs added to them. Owen maintained that in many real species of vertebrates some of the prongs were transformed into fins or limbs. In trying to lay out his Archetype theory, he spent much of his time trying to decide how he could produce the bones in the skull from different parts of a vertebra. It was esoteric work, and when he gave his first public lecture on his ideas, he didn't talk about it. Instead, he chose limbs. He knew that his audience would listen carefully, be-

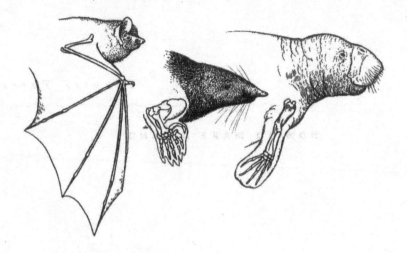

The homologous hands of a bat, mole, and dugong

cause we humans admire our hands—they are our instruments, to impose on the world what is in our minds. While other animals claw or climb or run with their distal pectoral appendages, we use ours to turn clay into pots, focus telescopes, paint gods. In a human, Owen declared, "one pair of limbs is expressly organized for locomotion and standing in the erect position, the other pair is left free to execute the manifold behests of his rational and inventive Will."

In limbs Owen also had a straightforward example of Cuvier's limits. He directed his audience's attention to the paddle of a dugong, the claws of a mole, and the membraned arms of a bat. One worked as an oar for swimming, another as a shovel for digging tunnels, and the third as a flying wing. Yet Owen could show how most of the bones in each limb corresponded to those in the others. Each animal had a single humerus at the shoulder, the radius and ulna coming off the elbow, a bracelet of bones at the wrist, and no more than five digits. Not for another 140 years would a creature be found with more than five. Often the order of the digits was preserved, even as bones might coalesce or disappear. Some fingers might be missing from salamanders, horses, or *Tyrannosaurus rex.* But the consistency among the remaining bones frequently made clear which wrist bone was missing its mate, showing which fingers of the archetypal five were lacking.

Cuvier had maintained that the bodies of animals were dictated by the kinds of lives they led. Every tetrapod limb is well suited for the animal's own life, and up to this point Owen granted that Cuvier was right. Yet they are all based on the same fundamental plan of a limb, and for this fact Cuvier had no explanation. Owen argued that our hands are actually the latest and most sublime manifestation of the Archetype. Fishes had only fins, but certain species such as lungfish were graced with limblike fins. Five-digited limbs first appeared on the primitive tetrapods. Later elaborations on the limb appeared on reptiles, birds, and mammals—but how God tacked them on his creations was a mystery to Owen.

Ten years later Darwin dragged Owen's Archetype down from the divine drafting table to the biosphere. The homology of limbs was a sign not of some ideal Archetype but of the fact that tetrapods descended from a common ancestor with legs and toes. Yet as brilliant as this insight was, it left Darwin with a dilemma that biologists have taken almost a century and a half to escape. Darwin argued that change is above all the product of generation after generation slowly adapting to local conditions, exploiting the variation built into reproduction to hit on new ways to survive. But once limbs evolved from fins, different lineages adapted them to swimming, burrowing, flying, climbing, and running, without ever abandoning the blueprint for building a limb. It's not that limbs are the only design that can work for each job: insects can fly with wings that look nothing like a bird's. Limbs are so similar, in fact, that natural selection alone isn't sufficient as an explanation. If toes are so good for giving an animal traction on land, then why isn't there a species with twenty toes? What's wrong with triple thighs?

The reason for the limits on limbs, biologists would ultimately realize, is bound up in the way macroevolution derived them from fins in the first place. Darwin himself had only a hunch of where to find enlightenment, but it was a good one: in the embryo. While he had been raising pigeons and thinking about how to create a new species, German biologists were recording how animals grew from a single cell. Species as diverse as a turtle and a pig all looked eerily similar as embryos. For a few days, what appear like the gill slits of a fish appear along the neck of a human embryo. The leading embryologist of the early nineteenth century, Karl von Baer, didn't think much of these similarities: he pointed out that Cuvier's four major groups of animals did not start like one another. A vertebrate resembled only other vertebrates after conception, and quite soon afterward fish and man went their separate ways from that common form.

Darwin snatched this antievolutionist work away from its creators as he had in so many other cases, arguing that evolution worked mainly by tinkering with the way the young of a species developed. The embryos of a pig and a turtle were so similar because they shared a relatively recent common ancestor, and as their lineage branched apart natural selection had made a series of changes to the way each developed. By tracing the growth of structures inside the embryos of different species, a biologist could see signs of evolutionary kinship that disappeared in adults. Most mammals have a separate premaxillary bone that sits at the tip of the snout, but humans do not. Yet as embryos we start out with a distinct premaxillary; by the third month it fuses with the rest of the upper jaw.

With Darwin's insight, embryology, which is the science of how embryos develop, can introduce us to some unlikely relatives. The sea squirt is a gorgeous blue sheath a few inches long that lives as an adult rooted to coral reefs, quietly filtering food out of the passing water like a respectable invertebrate. But it is actually born as a swimming, tadpole-shaped larva that cushions its major nerve axis with a long, gelatinous rope. As embryos we humans share this same structure—known as a notochord—but it is overtaken by the growing bones of our spine and soon dwindles to the gristly disks wedged between our vertebrae. Only when the sea squirt approaches adulthood does its notochord disappear and it roots itself to a reef. Thanks to Darwinian embryology, we recognize the sea squirt as one of the closest living relatives to vertebrates.

Folding embryology into evolution was one of Darwin's great accomplishments, but it also led to one of his biggest embarrassments. In 1866 it showed up in his mailbox, a thousand pages of feverish writing called *General Morphology*, composed by a thirty-two-year-old German named Ernst Haeckel. Germany had embraced the *Origin of Species*, and *Darwinismus* was practically a religion in some circles. Haeckel, a first-rate naturalist, was also one of its chief evangelists. Darwin had met him a few months before Haeckel's book came out, when he had come to visit Darwin at his country house. Darwin remembered him as a pleasant, cordial young man who spoke broken English. But as Darwin read Haeckel's book, his heart sank.

Life, Haeckel argued, was moving irresistibly onward and upward, and humanity was the closest to perfection it had achieved so far. Everything was part of life's evolution, civilization included, and natural selection only one force among many mysterious, unproven ones. Darwin puzzled over the new words Haeckel coined, such as *ontogeny*, which means the course of the growth of an embryo, and *phylogeny*, the evolutionary history of a species. In his book Haeckel put them together into one of biology's most infamous phrases: "ontogeny recapitulates phylogeny." By this he meant that evolution can only create new forms by adding extra stages to the end of an embryo's development.

He called it his Biogenetic Law: to make an amphibian, evolution added legs and removed gills from a fish embryo; a reptile or mammal was simply the product of still more added steps. As proof of the Biogenetic Law, Haeckel claimed you could actually see a shortened version of evolution in the growth of an embryo, from amoeba to invertebrate and upward.

Haeckel's ideas were wildly popular for a few decades, driving a number of paleontologists to try to find the ontogeny of today's animals in the fossils of their ancestors. Some thought, for instance, that evolution simply added extra steps of development to fins in order to create tetrapod limbs. Fins come in many shapes, but they are all variations on a basic form. A fin is made mostly of rays of dermal bone, the same stiff material that makes up scales, and deep in its core is a set of skeletal bones anchoring the muscles that move the fin. In many species, these skeletal bones are dominated by an axis—either a single long bone or a chain of smaller ones. Among lobe-fins, the axis takes over the fin bones completely, fitting into the shoulder like a tetrapod limb. The Brazilian lungfish's whiskerlike fin is an axis that has segmented into a series of bones. In the Australian lungfish's stouter fin, branches of bone radiate from both sides of the axis. And in Devonian lobe-fins such as *Eusthenopteron,* these branches come off only one side of the axis—the posterior side, meaning the one closest to its tail.

Paleontologists decided that in order to discover how fins had evolved into limbs, they had to figure out what had happened to the axis. The long bones of an arm clearly corresponded to the three big fin bones of *Eusthenopteron,* but the wrist and hand bones were trickier. A British paleontologist named D. M. S. Watson compared *Eusthenopteron* to early amphibians by inspecting the surfaces where each bone contacted its neighbors in a limb or fin. In that way he got an idea of how it was connected to other bones by ligaments and tendons and muscles; these connections might reflect how these amphibians grew. Watson decided that our hands were elaborated lobes, our wrist bones and fingers simply segmented branches. (Other scientists ran the axis through the little finger.) The earliest tetrapods had numerous wrist bones, and as their descendants evolved, they must have lost many of them, since we have only seven in our wrists and birds have three. In good Haecklian fashion, paleontologists thought that if you could look early enough in the development of a bird or any other living tetrapod, you could catch a glimpse of that primitive limb, which would then get pared down bone by bone to reach its modern form.

By the 1920s, however, Haeckel's laws were beginning to spontaneously combust. Critics assembled a bestiary of animals that defied Haeckel's rules. As adults, some species were embryonic versions of their ancestors; evolution sometimes added extra steps to the development of one organ while leaving the rest of an animal unchanged. Recapitulationists tried to put these cases down as exceptions, but by then evolutionary biologists were turning their attention to the new advances in genetics. By studying mutant breeds of animals, they could see that genes could insert changes almost anywhere along an embryo's development, and they looked back at Haeckel's ideas as merely quaint. Haeckel's fall was an embarrassment for embryologists, and they decided they would find only misery if they tried to use embryos to expand Darwinism. They stained their slides and mapped the fates of fetal cells, watching them as they huddled into organs or wandered across the body, and they avoided any speculation on what their results said about evolution.

Yet for all their flaws, Haeckel's ideas are worth salvaging. By some estimates, over 70 percent of evolutionary transformations from one species to another involve adding or subtracting a step at the end of a path of development. Such a figure may not be the result of an infallible Biogenetic Law, but it calls for an explanation. Only recently have researchers begun to reconsider Haeckel and at the same time think about how embryos can guide them toward a full theory of evolution. Much of the credit must go to British biologist Gavin de Beer in the middle of this century, and more recently to Stephen Jay Gould, who both ar-

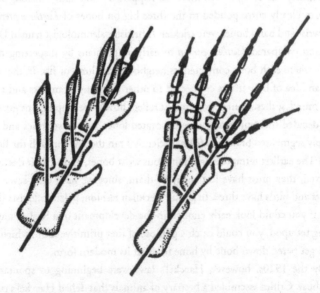

The axis as traditionally drawn through *Eusthenopteron*'s fin (left) and an early amphibian's hand. In this view, wrist bones and fingers are nothing more than segmented rays.

gued that Haeckel had recognized only one kind of change brought about by altering the timing of a growing embryo.

Gould put the idea most succinctly by asking us to think of an animal's development fixed not by a single timer but by a set of them. One timer may control its sexual maturity, while another controls the growth of its body. Resetting the timers, you can create a new organism. By the time an animal reaches sexual maturity and stops developing, it may have achieved only the body plan of a juvenile ancestor. That may be why the first baby chimps fooled early zoo-goers: as Owen showed, human heads are more like the heads of infant chimps than adult chimps. Alternatively, an animal may run through the entire development of its forerunners while it's still young and then simply continue the program, growing bigger horns, for instance, or more shell coils. Researchers now think that there are many different timers in a single animal, each controlling the growth of separate organs. Some salamanders, for instance, never manage to get out of their larval stage and live out their existence underwater. Yet the hemoglobin in their blood nevertheless goes through a metamorphosis into its adult form.

The seeds of a more surprising redemption of some of Haeckel's ideas came from the work of a mathematician named Alan Turing. Scientists who live on the harsh, lifeless plains of the physical sciences sometimes look at biology as a vacation spot—a lush green island they can visit, make a few groundbreaking discoveries, and then head back to the quantum steppes. After all, they say to themselves, if you know the laws of electrons and protons, if you can solve differential equations, you already know how Life works. Most of these scientists barely get off the plane before they discover that they were wrong—that biology's island paradise is a sweet-smelling swamp—and they either sink out of sight or catch the next flight out. But a few, such as Alan Turing, have managed to discover some original biological principles.

Turing, arguably the most influential mathematician of the twentieth century, is best known for exercises in logic that read more like dreams. In the 1930s he dreamed of a machine through which ran a strip of paper that stretched away on either side over the horizon and out to infinity. The tape was marked by a series of ones and zeroes that the machine could read, and depending on the number, the machine responded by moving the tape or changing the number. Turing showed that with only four rules such a machine could add. With a few more it could multiply, and Turing eventually proved that it could do any computation, given enough rules and tape. This logical exercise became the foundation on which computers were invented. We now live in the dream of this twenty-three-year-old mathematician.

During World War II the British army appropriated Turing's mind. He was in large part responsible for breaking the code systems of the Nazis, by simply thinking of them as Turing machines made real. After the war he helped lead the research that produced the world's first computers, but it was then that the tape of his life frayed. Since he was a teenager he had maintained a life of carefully hidden homosexuality. At age forty, though, he picked up a young man outside a movie theater, and a few days later he came home to find his house broken into, a few things missing—a compass, shoes, fish knives. He reported the boy's theft to the police, but when they realized that the burglar was his lover, they accused Turing of sex crimes. Rather than go to prison, he submitted to humiliating experimental hormone therapy and was exiled from his research on cryptography and computers. Two years later he was found dead in his bed, apparently having eaten a slice of apple he had dipped in cyanide.

In the last few years of his life, Turing gave a lot of thought to biology. How, he wondered, did patterns appear in nature? After a human egg was fertilized and divided a few times it became a uniform ball of cells. And yet on this unpatterned globe, a fold suddenly appeared along one side that would serve as the head-to-tail axis along which the embryo would grow. The cells of a growing starfish know how to organize themselves into a star; a leopard's skin can organize itself into spots. When Turing was thinking about such natural patterns, geneticists were still a few years away from discovering DNA, but they were already convinced that genes must control the unfolding of an embryo, probably by imposing a positional map on it with their proteins. Turing wasn't so sure. If the genes in a uniform ball of cells are all producing proteins, the proteins are so uniformly spread out that they can have no pattern themselves.

Embryologists in the early twentieth century tried to explain patterns with what they called a morphogenetic field. They envisioned that groups of embryonic cells might be caught in a biological analogy to electromagnetism. Before an eye has formed on an embryo, you can take the undifferentiated cells from that part of its head and graft them to another region of the body and an eye will still form. On the other hand, if you graft undifferentiated cells into a just-forming limb from elsewhere in the embryo, the foreign cells are overwhelmed by the morphogenetic field, forget their past, and help build a leg. Although geneticists scorned the morphogenetic field—they considered it virtually mystical—Turing realized he could create one with a few straightforward rules, as simple as computer codes.

He offered a hypothetical example. "This model will be a simplification and an idealization," he wrote, "and consequently a falsification." He asked his readers to imagine a ring of cells, in which every cell steadily produced two molecules (which we'll call Alpha and Beta). All of the cells are identical, and at the beginning of Turing's thought experiment the uniform levels of Alpha and Beta around the ring are at a steady equilibrium. Alpha spreads slowly from cell to cell around the ring, speeding up the production of more Alpha and Beta. Beta meanwhile inhibits cells from producing Alpha, and it can move around the ring much more quickly. With these properties, Turing showed that the initial equilibrium of the molecules is unstable, vanishing as easily as a stick standing on end can be toppled. All that's needed is a slight fluctuation in one cell's level of Alpha. If a cell churns out a little extra Alpha, Beta increases in turn and spreads quickly out into the surrounding cells like a wave, reducing the Alpha. With less Alpha in a cell, it makes less Beta, which allows Alpha to build up again. These feedbacks and fluctuations spread around the entire ring and eventually settle into a second, stable equilibrium: now the levels of Alpha and Beta form permanent waves around the ring. If they were pigments, the ring would be striped.

No divine brush or genetic map painted these stripes. The genes simply made a soup of interacting molecules that could be jostled into patterns, forming as easily as crystals in freezing water or harmonics on a plucked guitar string. With this elegant thought experiment Turing showed how life could find complexity for free, and mathematical biologists have carried his methods forward, simulating everything from the designs on shells to the stripes on zebras and the spiral patterns of sea urchin embryos. And in the early 1980s some scientists began to wonder if Turing patterns might be shaping limbs.

An embryo of a human, a chicken, or a salamander starts off as a limbless, curled tube of tissue. Its notochord is still huge at this point, running alongside a digestive system that is still only a cylinder of gut with a few ducts divoting its surface. Its brain is a stump, its eyes goggly. When the embryo is twenty-four days old, the cells in the region of newly formed kidneys release a signal that triggers other cells in the embryo's flanks to gather on the surface of the embryo. They congregate and swell like blisters into limb buds. Inside the buds, cells known as mesenchyme cells nestle in a matrix of jelly. A lip of tissue forms on the edge of the bud (called the apical ectodermal ridge) and stimulates the mesenchyme cells to multiply. Take away the ridge, and no limb forms; add a graft of ridge, and you can get extra fingers or arms when a mouse is born. Only the mesenchyme cells close to the ridge proliferate, building on the existing bud and extending it away from the body, forming it into a cylinder—what will become the humerus.

After a week some mesenchyme cells in the bud draw toward one another in the shape of the future limb. As they cluster together, they change into cartilage, oozing out connective tissues that form a soft mold. Meanwhile blood vessels have been snaking their own way into the limb and now slip into the center of these cartilage casts. They deliver bone-generating cells that fill the forms that the cartilage cells defined, while most of the cartilage dies away. Muscle and tendons cling to the bones and stretch as the bones grow. As the hand flattens and rounds out, many cells die, the total suicide count depending on how free the digits will be in the adult. In the hand of a human embryo the tissue dies back to the base of the digits, leaving them free to thread needles and play trumpets in later life. A duck's feet are more merciful, letting the tissue remain to form webbed paddles.

It's not too odd that mesenchyme cells might gather into a single cylinder, but how could they possibly know to branch at the elbow into a radius and ulna, and later consolidate into wrist bones and fingers? Perhaps the anatomy wasn't encoded in genes so much as the natural result of the shape of the limb bud and the interaction between neighboring mesenchyme cells. Some researchers looked for morphogenetic molecules in the stew of chemicals that mesenchyme cells secrete. Others were struck by the way that the cells themselves crawled and crashed together. A mesenchyme cell creeps through the goo inside a limb bud like the Blob moving through a tub of marmalade. The cell stiffens parts of its membrane into tentacles that reach out and grab hold of anything they encounter. As it hoists itself forward it tugs on the surrounding matrix, deforming it so much that a nearby cell may find itself pulled toward the crawler. Chances are good then that the cells will touch each other with their tentacles, and since mesenchyme cells have sticky, sugary coatings, they will cling together. When enough cells have condensed together, a chemical drips out of their pores and dissolves the marmalade in its vicinity, making the surrounding cells collapse in on themselves more.

The force of thousands of cells jostling and tugging on the matrix of the limb bud combine to change the shape of the entire bud. Like Turing's chemicals, the forces create a morphogenetic field. When a group of biologists led by George Oster at the University of California at Berkeley modeled these forces, they found that they spontaneously created something that looked like a growing limb. When a simulated bud emerged from the wall of the embryo, the mesenchyme cells gather into a single disk. As the cells near the ridge multiply, the disk elongates into a cylinder—what will become the humerus. But Oster's model also shows that the traction from the tugging of the cells flattens the entire limb bud, which in turn flattens the cylinder of mesenchyme cells itself. As the cluster flattens, the pull of the mesenchyme cells toward one another makes the cylinder split in two, like drops on a car hood separated by surface tension. They break off into two smaller cylinders, corresponding to the radius and ulna, and as they in turn grow, they flatten out as well. Beyond a critical threshold, the two cylinders split into still smaller clusters that vaguely resemble wrist and fingers.

Oster's models inspired one of his former students, Pere Alberch, to see if he could find any evidence of a real morphogenetic field in the limbs of living animals. While working at Harvard in 1983, he dabbed newly formed limb buds of frogs with a chemical that slowed down the rate at which the mesenchyme cells divided. He found that with fewer cells in their limb buds the frogs didn't grow a smaller version of a normal hand; instead they usually lost their thumb completely. Since Alberch hadn't sabotaged the DNA of these animals, he couldn't explain his results with genes alone. A morphogenetic field like Oster's might make more sense, since changing the size of a limb bud could have altered the forces that determined how it branched—much as the way you change notes on a slide whistle by changing the size of the volume in which air vibrates.

After he had finished with these experiments, Alberch even stumbled across a natural experiment when he went back to Spain to spend some time with his parents. They showed off their new Great Pyrenees dog, and naturally Alberch immediately checked its toes. Like wolves, most dogs have four on their front feet, along with a vestigial fifth. Smaller dogs like poodles rarely have more than four, while bigger dogs, like Saint Bernards, sometimes even have a sixth toe, despite the best efforts of breeders to wipe it out. The Alberchs' even bigger Great Pyrenees had six, which Alberch learned is standard for the breed. People have suggested that the sixth toe is somehow an adaptation in these dogs for walking through mountain snow, but Alberch's experiments make him think that it's the result of the vagaries of development instead. The genes of each breed probably don't carry with them a certain number of digits. The size of the limb bud instead influences how many toes will form.

Alberch had also been deeply influenced by the work of Stephen Jay Gould, and so he wasn't satisfied with simply explaining how a given limb developed: he wanted to know how evolution had engineered it in the first place. He was well aware that the old Haeckelian idea that limbs evolved through a simple, bone-by-bone recapitulation of phylogeny by ontogeny was wrong. The harshest refutation came from experiments in the 1970s in which researchers were able to study chicken wings and feet for the first time during the earliest stages of the condensation of their cartilage. These limbs did not start out as some kind of primitive amphibian leg: from their first appearance, the mesenchyme cells immediately clustered into patterns close to the ones that the bones would take. Another problem had to do with homology. Haeckel (and Darwin and Owen before him) thought it was simple enough to find a correspondence between the bones of one tetrapod limb and any other. But the longer embryologists looked at limbs, the less sure they became. They could not say whether the three fingers of a bird wing were 1, 2, and 3 or 2, 3, and 4. Wrists and ankles were even more baffling. Haeckel would have wept.

Alberch wondered if there might be a deeper order below this confusion. He set up a lab at Harvard where graduate students could study the day-by-day growth of limbs in tetrapods such as frogs, turtles, and alligators. And one day while he was telling a class about the puzzles of limb evolution and development, he grabbed the attention of a twenty-four-year-old paleontologist named Neil Shubin.

Shubin had been laying the groundwork for a promising, albeit conventional, career as a paleontologist. He had been spending his summers along the Bay of Fundy on the northern shores of Nova Scotia. Where the ocean meets the land, wide tidal flats stretch from the water to high cliffs that face the Atlantic bare, the rock dating back between 175 and 225 million years. Shubin hoped to find fossils at the end of an inlet, a place older scientists had declared pointless to explore, both barren and dangerous. Every day the tides in the bay would rise as much as fifty feet, and if a person lingered too long on a ledge, the flats would vanish and the cliffs would cut off any retreat. Shubin and his fellow paleontologists took the risk anyway and discovered that the cliffs were actually speckled with hundreds of thousands of bone fragments. The fossils were of some of the closest relatives to the first mammals, of galloping crocodiles smaller than beagles, of dinosaurs the size of sparrows.

The lode could have set Shubin's life: he could have spent his career doing nothing else but mining the fossils and reconstructing this critical stage of life's history. But he liked to be distracted. While talking to Shubin you can get intellectual whiplash as he turns the conversation from one field to another, from geochronology to genetics, from the intricacies of mammal teeth to the biomechanics of a jumping frog. When he wasn't racing tides on Nova Scotia he was taking classes at Harvard, and from time to time he would distract himself by going to the rare book room at the library. There he would read the works of old masters, such as Owen's 1849 *On the Nature of Limbs*. It was almost painful to read the exquisite descriptions of limbs, which were followed not by a Darwinian epiphany but by a retreat to the notion of an Archetype. "I got into that, just the elegance of the argument," Shubin says, "and the fact that he was coming right up against evolution—right up against it—and then backing off and doing a round-end on it with this transcendental notion. It's just amazing, here's this guy who was an excellent anatomist, and when he's talking about similarities in structure, he came so close."

Alberch was helping to carry on this deep tradition of studying limbs, and so Shubin had long talks with him about embryos and fossils and evolution. Eventually Shubin decided to see for himself what connections there might be between ontogeny and evolution by embarking on a tour of every tetrapod limb he could study. "People in Alberch's lab were collecting a lot of developmental series of vertebrates, and there was a great library of Harvard," he explains. "I pulled out every paper I could possibly find on any group—didn't make a difference. I made these huge stacks—birds, mammals, and so on, and I went through them." He talked to Alberch's students about how the limbs of alligators, turtles, and other tetrapods formed, and he did research himself on salamanders, flushing their embryos clear with alcohol and glycerine and then staining their cartilage blue, or plunging the embryos into plastic blocks, scraping off layer after layer, and using a camera lucida to draw the four-fingered hands at each stage of their development.

When he looked over all the information he had collected, Shubin didn't quite know at first how to make sense of it all. He certainly didn't see any strict sort of recapitulation in the growing limbs, and yet he thought he was catching glimpses of an underlying unity in the way they formed. In many tetrapods he noticed that the mesenchyme cells in the wrists and ankles condensed along an arch that reached day by day from one side to the other. In some animals the arch disappeared as clusters fused together, but in others it lasted through the tetrapod's life. We humans grow an arch in our wrists, and you can see in *Gray's Anatomy* how nerves and blood vessels follow along its path. But the arch wasn't an absolute rule, since some tetrapods never formed it.

As Shubin and Alberch pored over the results, it occurred to them to think back to Oster's model. Oster's simulated limbs always formed through a few kinds of events: mesenchyme cells first clumped together into clusters, these clusters then sometimes split into two branches, and they also sometimes merged or split into segments. Alberch and Shubin realized that in the sequences of actual developing limbs, every one formed by the same short chain of events. They could sketch out the growth with a set of symbols, writing down the development of a limb like a sentence. The illustration on page 70 shows the story of a mammal hand. The humerus forms first, stretching until the cells thin out, and then branches into the radius and ulna. The growth along the radius continues into one more bone, a small wrist bone called the radiale (which in humans fuses back into a knob at the end of the radius). Meanwhile the ulna branches into two wrist bones, which branch in turn into more bones. One branch extends across the top of the limb, forming the digital arch, and the fingers sprout from it.

As different as tetrapod limbs may look in adult animals, they all follow these rules as embryos. In a frog hand, with only four fingers, the arch simply stops growing before a fifth digit can branch away. The olm, a European relative of salamanders, has only two toes, but it never even develops the other three. As the olm embryo develops, only two small bones form in its ankle, and from one of them only two toes branch. At the same time, a branching, bending axis

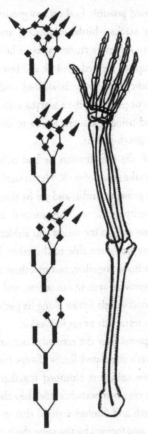

Tetrapod limbs begin growing as a branching axis of mesenchyme cells, which condense into cartilage that is replaced by bone. Rather than extending through a finger, the axis hooks across our wrists.

might explain the results Alberch had gotten in his experimental frogs: the thumbs disappeared because they branched off at the end of the hook. In these patterns, Shubin and Alberch had encountered a new kind of homology: not Owen's bone-by-bone correspondence, but a homology of growth. A homology of growth explains why you never see triple thighs: the branching rule makes them mathematically impossible to evolve.

As a paleontologist, Shubin recognized that these patterns also pointed to a new explanation for how tetrapod hands evolved from fins in the first place. Our ancestors' fins were built from an axis with branches along one side. Tetrapods retained this anatomy—our radius and a few wrist bones still branch off the same side—but as the first tetrapods evolved, the axis didn't extend straight into a finger. It took an abrupt turn instead near its end, hooking across what became the wrist. The branches flipped to the other side of the axis, and formed fingers. In other words, Haeckel was—to a limited extent—right: tetrapod arms really are an extension added onto lobe fins, and in those brief days before the hand begins to form, our arms probably do bear a faint resemblance to an embryonic *Eusthenopteron*. Did a new morphogenetic field create the hook? Did a new system of genes take over? Shubin and Alberch couldn't say. All they could do was point out the dynamic patterns and hope for the discovery of the underlying mechanism.

They published their findings in 1986, producing what Shubin describes as "an enormous thud." No one wrote him a letter asking about the data he had found; no one cited him. The science of embryos and the science of fossils had drifted so far apart that few were interested in what clumps of cartilage could say about the biggest transition in vertebrate history. Shubin shrugged and went on with his life. At Berkeley and then as a professor at the University of Pennsylvania, he studied salamander hands and searched for more early mammal fossils in Nova Scotia, Morocco, and Greenland. He had said his piece.

But Mike Coates knew about the new hook. When he came to Cambridge two years later and put himself through a crash course in tetrapods, he read about it. And as *Acanthostega* opened its eight-fingered hand to him, he was glad he did. The old theories of how fins became limbs assumed that the first tetrapods had five fingers, and drew branches from the limb axis through the wrist bones and fingers accordingly. Now Coates had found three extra fingers on an early tetrapod, and there were no wrist bones left through which to draw even one extra branch. If evolution had built fingers on Shubin's twisted arch, however, the matter was far simpler. You could grow five fingers along the bent axis—or six, seven, or eight—by simply hanging them from the end of the hook. And in time other Devonian tetrapods turned out not to follow the five-finger rule. *Tulerpeton*, the mysterious Soviet creature, had six spindly fingers, and Coates discovered that the *Ichthyostega* leg that Jenny Clack had found in Greenland had seven toes. Our five-fingered hand was a low-digited variation among many, and only Shubin and Alberch could accommodate them all.

The idea that Turing patterns are molding limbs has its share of detractors. According to the model that Oster had built, the forces acting on the limb bud as a whole generated the pattern of condensations that later became bone—not any sort of earlier chemical signaling that might map out the future shape of the limb, when the bud was still a uniform batch of cells. In 1990 English biologists set up an experiment to test the two possibilities. They snipped off half of the bud of a chick wing, and in its place they grafted the other half from a second wing. They made the switch early in the bud's growth, when to the naked microscope it looked like nothing but a uniform porridge of mesenchyme cells. In Oster's model such a bud would still produce a normal arm, because the morphogenetic field had not yet come into play that would shape the cartilage. But the experimental buds produced double humeri. The results led the biologists to conclude that invisible markers must have already been laid out in the limb bud and its graft before the experimental exchange was made, and long before Oster's forces could have had any affect.

It was impossible to weigh alternatives, though, because at the time geneticists hardly understood how genes construct an embryo. For decades they had been contenting themselves with studying mutants. One set of freaks, first noticed by a British zoologist named William Bateson in the 1890s, was particularly striking. A sawfly was born with feet on its antenna. A spiny lobster had an antenna where its eye should have been. Moths had wings for legs. Mammals had similar, though subtler, sorts of mutations, such as neck vertebrae that sprouted little ribs. Bateson named them homeotic mutations for the way body parts retained their shape despite their change of address.

It was only in the 1980s that geneticists were able to trace homeotic shuffles of anatomy to homeotic genes at work in a developing fruit fly larva. When they had identified all of the genes, they stepped back and were amazed at the simplicity of what they had found. The homeotic genes are strung up on a fruit fly's DNA like a chain of lanterns; the first gene in the chain switches on and makes proteins in some of the cells of the larva's head, and the rest activate in corresponding order all the way back to its tail. The larva is made up of segments,

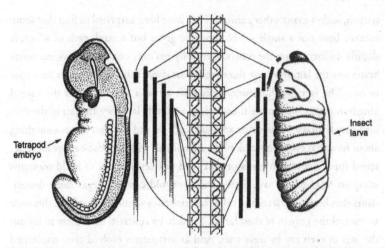

Nearly identical homeotic genes (represented by boxes) determine the fate of segments of an insect larva as well as those of a tetrapod embryo. The bars along each embryo show the range in which cells express a given gene to make a protein.

and in each segment a unique combination of overlapping homeotic genes churns out proteins in the cells. These proteins travel to other regions of the DNA in each cell, where they turn on other genes that then produce the anatomy that each segment needs, be it a leg, a wing, or an antenna. Bateson's homeotic mutants were the result of homeotic genes that switched on in the wrong segment, giving it a new identity.

Vertebrates follow a profoundly different developmental path than a fly, which isn't surprising given that their last common ancestor lived about a billion years ago. Vertebrate embryos also have segments running down their backs, called somites, but somites develop into internal skeletons—spines and ribs—as well as skin and muscle and nerves. But in the late 1980s geneticists discovered that there were genes at work in chick and mouse embryos that were almost identical to the homeotic genes in insects. A homeotic gene from a human can be put in a fruit fly's DNA and do the job of its counterpart. Vertebrate homeotic genes also mapped out the head-to-tail axis of embryos, and if geneticists created mutations in these genes, they got homeotic mutations. Neck vertebrae took on the shape of vertebrae behind the rib cage. The second gill-like pouch in a mammal embryo normally produces cells that ultimately form the stapes, one of the bones of the middle ear. But if the homeotic genes are toyed with, the pouch produces an incus and malleus instead—the other two bones of the ear, which the first gill pouch usually creates.

Within a few years, scientists found the same genes in worms, in leeches, in starfish, and in many other animals. They have been surprised to find that some animals have not a single set of homeotic genes but several, each of which is slightly different from the others. A fruit fly has only one set, but jawless vertebrates like the lamprey have three, and most jawed fish and tetrapods have four or five. The fact that homeotic genes exist across a wide swath of the animal kingdom makes them more than tools in a fruit fly lab: they are part of the evolutionary heritage of the entire animal kingdom, and they can tell us something about how that evolution took place. All animals with homeotic genes must descend from a common ancestor that used one set of them to build segments along its head-to-tail axis. This billion-year-old creature begat some descendants that became insects and other invertebrates—animals that used this code to control the growth of their larvae. Starfish, by contrast, used them to lay out the axis of each ray in their stars. And as vertebrates evolved they duplicated their homeotic genes several times; in one case around the time the first vertebrates arose, and later when the first vertebrates with jaws appeared.

The duplication of an entire set of genes is a pretty radical mutation, and yet on a scale of millions of years, these mutations happen often. At first a duplicated gene simply makes the same proteins as the original. But the mutations that are continually attacking DNA have an easier time altering a duplicated gene because a change to one gene doesn't affect its twin, which can continue its old job. For the most part these extra copies simply burden DNA with useless extra genes, but sometimes a duplicated gene can mutate into a new form that performs a new task.

Immune cells in many animals, for example, use proteins called lysozymes to fight against bacteria. The cells engulf a microbe and unleash their lysozymes, which latch on to its surface and drill holes. Molecules leak in and out of the bacteria's membrane until it dies. In humans, as in most mammals, the only place where lysozymes exist outside of the blood is in tears, where they protect the eyes from infection. But ruminants—cud-chewing, hoofed mammals such as cows, goats, and sheep—are unusual for having many mutant lysozymes in their stomachs. An ancestor of the ruminants apparently duplicated the gene that produces lysozymes, and the second set became adapted to the harsh acid of its stomach. Ruminants have a symbiotic relationship with bacteria that live in a chamber where they digest tough plant matter, and the ruminants periodically swallow some of the microbes into their stomach. There the new form of lysozymes can drill open the microbes, freeing up the vitamins and nutrients that the ruminant can digest.

The duplication of lysozyme genes made an enormous difference in the evolution of ruminants. But the duplications of homeotic genes—which are so critical to establishing the body plan of embryos—was a far more important macroevolutionary event. A duplication of Hox genes (as the vertebrate forms of homeotic genes are called) appears to have coincided with the origin of the head—a bony shell around a cluster of nerve cells equipped with two light-sensitive organs. A second duplication occurred when vertebrates evolved a jaw from some of their gill arches, beginning to transform themselves from minnow-sized muck-suckers to marine predators, and within 100 million years into enormous hunters that could stave in a boat. Each of these changes demanded a more sophisticated symphony of development, and a duplicated set of Hox genes, glumly cranking out superfluous proteins for laying out a primitive body plan, could be easily drummed into service to help build new structures.

In 1989, as biologists were mapping the web of interacting proteins that are created in a growing limb bud, they began to realize that Hox genes are at work there as well. To understand this network, you have to get an image of your arms as they first formed, with the thumb side pointed toward your head and the little-finger side oriented down toward your feet. A region of the limb bud along the little-finger side known as the Zone of Polarizing Activity secretes a protein that a researcher named after his favorite video-game character, Sonic Hedgehog. Sonic Hedgehog proteins ooze from the zone to the ridge of the limb bud, where they trigger the ridge cells to secrete a growth-stimulating protein, which travels in turn to the neighboring mesenchyme cells and makes them divide like mad. At the same time the growth-stimulating protein creates a feedback by encouraging the Zone of Polarizing Activity to produce more Sonic Hedgehog, which goes back to the ridge. And while this frenzied volley is going on, both of these proteins also shower the mesenchyme cells and signal them to switch on their Hox genes.

When scientists first detected Hox genes at work in limbs, they speculated that they might work as they did along the spine: the limb would be divided like a sliced loaf of bread, and a unique combination of Hox genes in each slice would sculpt the bones to the proper shape. The head-to-tail axis, they imagined, had simply been turned into shoulder-to-hand. The scientists tampered with the Hox code in the limbs of mice and chicks and waited to see if the mutants confirmed their suspicions. The early experiments seemed to. When geneticists knocked out the last gene of one set of Hox genes, for example, mice were born with a shrunken tail and withered fingertips.

But more experiments showed that reality was messier than predictions: another set of Hox genes switches on only along the little-finger side of the limb bud and then takes a sharp turn to spread over the fingertip edge. Confusion reigned. Fortunately, though, Coates was always on the lookout for new work that could help him understand how vertebrates evolved hands, and he kept tabs on the Hox papers as they were published. When he saw the swing of the Hox genes, he knew he had seen the pattern before: it was the curve of Shubin's digital arch. "I looked at it," says Coates, "and thought, 'This is unbelievably nice.'"

Coates brought this serendipity to the attention of the scientific community, showing them how fossils, cells, and genes were all hinting at the same explanation for this case of macroevolution. No one knew enough about the limb to say how Hox genes helped form Shubin's digital arch, but whatever the mechanism, the correspondences were too clear to ignore. Inspired by Coates, Swiss biolo-

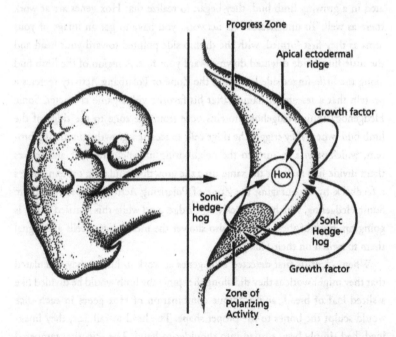

A network of genes, a few of which are shown here, produces the limb. Sonic hedgehog in the Zone of Polarizing Activity induces the apical ectodermal ridge to produce a growth factor, which in turn stimulates mesenchyme cells to multiply. Their feedback makes the limb bud grow, and together they guide the expression of Hox genes in the Progress Zone. Hox genes play a major role in determining the limb's structure.

gists demonstrated in 1995 that Hox genes also make fish fins. The species they chose was the zebra fish, a little ray-finned fish that, like a mouse, is well suited to life in a lab. Zebra fish have a fast breeding cycle, a quick growth into fry, and most important, a transparent body in early life that makes it easy to see protein-sensitive stains in their innards. The Swiss found that some of the same key genes that form a human arm form the front fin of zebra fish, and in a pattern that's practically identical at first. A clump of mesenchyme pimples up along a zebra fish flank, and a ridge then appears along its outer edge. A Zone of Polarizing Activity in the tailward position begins to brew its broth of Sonic Hedgehog proteins; ridge and zone speak to one another; and in response one set of Hox genes begins to produce its proteins along the tailward edge. The mesenchyme cells start to clump together into clusters.

If zebra fishes were tetrapods, the cells along the far edge of the fin would then start to produce Hox proteins, creating the tetrapod hook. Instead the Hox genes simply shut off. The mesenchyme cells condense along the straight axis marked by Hox genes, and then divide into bones that all attach directly to the shoulder. But although the Hox genes stop working, the fin continues to grow. In tetrapods the ridge of the limb bud fades away once it has stimulated the mesenchyme cells to multiply, but in the zebra fish the ridge itself begins to grow, stretching out into a great flat plate in which fin rays develop.

These results agreed with the predictions of paleontologists so well as to be a little creepy. The common ancestor of ray-finned fish and lobe-fins probably had a modest fin. Its Hox genes formed an axis for the skeletal bones to form; small branches grew from the axis, and beyond them the ridge produced a small fin. Its descendants formed two great branches, along which the timing of development sped up and slowed down. Along the branch that led to ray-finned fishes, such as zebra fish, the Hox genes switched off earlier, while the genes in the ridge turned on sooner, producing bigger fin rays. The lobe-fins meanwhile let their Hox genes work longer, producing a longer axis and a bigger set of skeletal bones. At the same time, they delayed the growth of their fin rays, so that they became a fringe surrounding the internal skeleton. And the lobe-fins that led to tetrapods took this change in timing even further: the ridge of the limb bud stopped growing so early that no fin developed at all, while the axis hooked around the far edge of the limb bud, giving rise to fingers.

In 1996 an American group of geneticists answered the Swiss with a detailed portrait of the developing tetrapod limb. Following the progress of twenty-three Hox genes day by day, the scientists saw that they corresponded to Shubin and

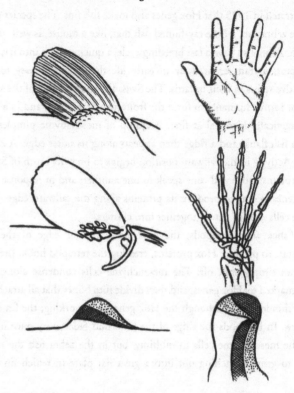

In a zebra fish (left), the parts of the fin where Hox genes are active mark a straight axis, along which bones later grow. In tetrapods, the same genes hook around the end of the limb bud, and in this new region they flip their order. This pattern matches the later development of the limb axis. The branching takes place on the thumb side of the limb until it reaches the wrist, and at that point the axis bends and the branching flips to the other side, becoming fingers.

Alberch's hook more closely than anyone had imagined. In the early stages of a limb bud the regions where Hox genes were active formed stripes running up the little-finger edge. After the bud had grown larger, the stripes ran along the far edge where the fingers would form—but the stripes reversed their order. Back when Shubin and Alberch were transforming lobe fins into tetrapod hands, before anyone even knew that Hox genes had something to do with them, they had seen this flip. In an animal like *Eusthenopteron,* all of the skeletal branches extend from the axis toward the thumb side. When tetrapods evolved their hook, the branching switched to the other side of the axis, where it created fingers. To flip the branching, it now seems, evolution must have flipped the Hox genes.

Shubin spent years thinking that his work on the bent axis would be a minor footnote in his life's work, but now geneticists are roping him into coauthoring papers about limb development. While they can read sequences of genes and stained cells, he can read the code of fossils, which holds information no living animal can offer. Recently, for instance, he and Ted Daeschler have been working on a fossil that Daeschler found in 1996 that may be a fin with fingers. The fossil belonged to a Devonian lobe-fin called *Sauripterus*. It was the first lobe-fin found in the Catskill Formation, but before Daeschler's discovery, the best fossils were badly beaten up. As far as paleontologists could tell from them, *Sauripterus* belonged to a group of extinct lobe-fins that were closer to *Eusthenopteron* and tetrapods than they were to lungfish, but almost definitely not the lineage that actually gave rise to the terrestrial pioneers.

Daeschler's fossil preserves the fin from its shoulder to the tip of its rays. You can stroke the smooth scoops where muscles attached to the bone, and you can see rounded joints where they could bend. *Sauripterus* apparently could use massive chest muscles and triceplike muscles in its fins to prop up the front of its body. It had big bones corresponding to our humerus, radius, and ulna, and beyond these it had at least eight slender rods of skeletal bone that are like the bones in our own hands. Although these fingerlike bones were still swaddled inside the fin, *Sauripterus* could bend its fin at the equivalent of a wrist, as well as at the joints of its fingers.

Scientists are now in a position to look at such a fossil and recognize the extinct genes that were at work in it. Clifford Tabin, a Hox gene expert from Harvard Medical School who collaborates with Shubin, has pointed out that close to the shoulder *Sauripterus*'s bones branch off the axis toward the "thumb" side of the fin, as in *Eusthenopteron* and other lobe-fins. But at the end of the fin, where the fingerlike bones form, some of the branching reverses direction, reaching instead toward the "little-finger" side. This was not a tetrapod hand by any means—*Sauripterus* never put a hook in its limb axis. Nevertheless, the fossil shows that this lobe-fin was trying out the same genetic recipe that our own ancestors were. Tetrapods were part of a widespread evolutionary experiment in handlike structures, but they happened to be the only ones to survive past the Devonian.

Evolutionary biologists like to think about hands because they are not some minor adjustment to an old structure; they are clear-cut innovations. For 3.5 billion years nothing on earth had hands. In a short period of time they came into existence, and they have remained pretty much unchanged since then. When

William Bateson first discovered homeotic mutants he thought he had found the secret to newness. Bateson was a powerful thinker—he was one of the first biologists to recognize that the work of an obscure nineteenth-century monk named Gregor Mendel contained clues to how heredity worked. He then coined the name *genetics* for the field he proceeded to help found. Bateson was obsessed with the mystery of variation—the coal that fed Darwin's locomotive of natural selection—and in studying hundreds of examples, he decided that most variation could lead only to relatively minor changes in colors, shapes, sizes. Of Darwin, the champion of minor variations, Bateson said, "we read his scheme of Evolution as we would those of Lucretius or of Lamarck." Only radical variations such as homeotic mutations, Bateson decided, were the truly creative powers in evolution, forces that could suddenly manufacture new things such as a tetrapod limb. The biologist Richard Goldschmidt later named these hypothetical creatures "hopeful monsters." They lurched into the world, drastically different from their ancestors, and if they were lucky they passed through natural selection's filter.

The science that Bateson founded, however, turned against him. There is no fundamental difference between a homeotic mutation and the kind of small variations that set the color of hair or eyes. The only reason that Hox mutations are so drastic is that the altered genes happen to regulate many other genes involved in building an embryo. While Bateson thought that homeotic mutations were necessary for creating a new species, the slow accumulation of small ones will actually suffice. Neodarwinists declared that new structures were similarly the result of gradual changes as natural selection worked on normal variations. In this respect, they argued, macroevolution didn't need a separate explanation from microevolution: it was simply microevolution left running for a few million years.

Both sides of this debate took up their positions with only a crude knowledge of how embryos form. With so much more information, a growing number of biologists now think that the either-or choice between gradual increments or giant mutations is a false one. At the same time, though, they believe that innovations are in fact distinct macroevolutionary events. A leading advocate of this shifting paradigm is Guenter Wagner. A Yale biologist who considers himself a card-carrying neodarwinist fully versed in population genetics, he thinks that innovations call for a fundamentally different explanation from that of more mundane kinds of change. "I think there is a break," he says.

To some extent, Wagner thinks Bateson was right: natural selection often sculpts rather than creates. "If you look at an organism," he explains, "let's say the Darwin finches, and you look at evolutionary change that can be understood as adaptation, they are usually modification of a given design. The beak, you must make it bigger, stronger. But there are other transitions that are contingent on a certain situation that is not available readily in any wide population." And yet to work through these transitions where something never seen before comes into existence, Wagner doesn't have to depend on some mysterious macromutation. The normal replication of genes can, for instance, accidentally spew out a duplicated gene, which natural selection can later adapt to a new task. Even more common are cases in which natural selection, focusing its attention on normal variations in shape and size, inadvertently alters the program by which an embryo develops, triggering a new cascade or possibly erecting a new morphogenetic field, so that a crude new structure forms.

Biologists have in recent years been uncovering small but telling examples of how this derailing can happen. Gerd Müller of the University of Austria has studied how the peculiar shin of birds forms. Below the knee of all tetrapods are two long bones, the tibia and the fibula. On birds, the tibia runs from knee to ankle, while the fibula is a sliver that clings to the tibia's upper half. A bird tibia has a unique blocky crest against which the fibula rests, glued tightly in place by connective tissue. Without the crest, a bird would have a difficult time walking, for a large muscle from the thigh fastens to the splinter-shaped fibula. Without a tight link from fibula to tibia, the muscle's force could never be transmitted down to the foot.

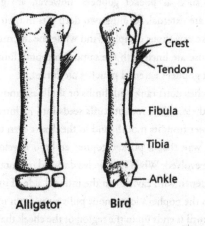

Crest

Tendon

Fibula

Tibia

Ankle

Alligator   **Bird**

An alligator shin and a bird shin with its evolutionary innovation of a crest on its tibia.

As essential as the crest is, no genes in a bird's DNA carry a specific line of instruction that summons it into existence. When a bird's leg forms, the two shin bones are of equal length, and the tibia lacks a crest. Instead, the tendon that attaches the thigh muscle to the fibula extends all the way to the tibia. In the days that follow, the tibia grows quickly as the fibula lags behind. Birds are restless in their eggs, squirming around when they are only a week old, and in the process they push and pull the fibula. It's a general rule that pressure will turn developing tendon tissue into cartilage, and so the bird's fidgeting automatically stimulates a nodule of cartilage to form between the shin bones. When the bird hatches and walks on its wobbly legs, the nodule fuses to the tibia and turns to a bony crest. If a bird cannot squirm in its egg, no crest forms.

Müller has found the same crest on fossils of the bipedal carnivorous dinosaurs from which birds descended. The early dinosaurs and their close relatives may have started to slim down their fibula in order to develop a swift run, and as a result the bone became unstable while it was still in its egg. In response, natural selection guided the tendon to the tibia to hold it steady, and as the fibula continued to shrink over millions of years, fidgeting embryonic dinosaurs put more stress on it until a crest appeared. The fibula was able to attach firmly to the crest, and so anchored, it could dwindle down to a splint.

Another example Wagner points to is cheek pouches. When many rodents forage, they pack seeds into cavities in the internal lining of their cheeks. They can then bring a lot of seeds home in each trip and store them away rather than make extra rounds under the watchful eyes of owls and rattlesnakes. A few families of rodents such as pocket gophers, however, are graced with deluxe pouches. Theirs are external, forming two deep pits in their cheeks with openings that look like cavernous dimples. And while the internal cheek pouches of a squirrel or mouse are lined with the same damp pink lining of the rest of its mouth, a pocket gopher's external pouches are furred.

External pouches don't rank with limbs or skulls as innovations go, but they are useful nevertheless. As a rodent stuffs seeds into its internal pouches, it inevitably loses water from its mouth, and in the desert even that loss can make a difference. The way that a pocket gopher embryo develops shows how the pouch may have evolved. When it is twelve days old, its mouth looks much like that of other rodents, with cavities on the internal side of its cheeks. But in the next few days as the gopher's long snout pulls away from its face, it carries the cavity forward until it ends up in the region of the cheek that will ultimately become the gopher's lips. The lips roll out of its mouth and onto its face, pulling the pouches forward with them. Now they no longer open to the interior of its mouth but out to the air. And as soon as they do, they make contact with the cells that are forming the skin on the gopher's face. The face is awash in signaling molecules that trigger tissue to sprout fur, and the cheek pouches obey the command and turn hairy. It's likely that the ancestors of pocket gophers—rodents with internal pouches—evolved a longer snout, and as a side effect this innovation came to be. In a sense, its evolution was both gradual *and* sudden: the gopher's snout grew slowly over the course of many generations, but there was no intermediate stage between the internal and external pouch. It went immediately from one form to the other.

On the other hand, the recent discoveries of how Hox genes work have made Wagner wonder if at least a few innovations did burst into existence all at once. What made him change his mind was work he and some of his students did on frogs. Frogs have a rare sort of ankle bone: it is long and shin-shaped rather than round, like a whittled marble. This innovation allowed frogs to generate more leverage in their legs for leaping, but it is also far beyond the bounds of ordinary variation in tetrapods. In most tetrapods, ankle bones may be slightly bigger or smaller, but under the sway of genes like Hox, they remain basically marbles. Frog ankle bones do not begin to form as little nodes of mesenchyme cells that gradually lengthen, but are long rods from the start. "That's a very nice innovation, because now they have an opportunity to grow more than any genetic variation on the size and shape of a nodular element can have," says Wagner.

Wagner has been studying how Hox genes build frog legs, focusing on one gene that normally plays only the role of shaping long bones such as the thigh and shin. But he found that in frogs, unlike any other tetrapod yet studied, this particular Hox gene also works its magic on the ankle as well. "If this is correct, it's a one-step process, it's not a gradual shape change. If it were a gradual shape change, you would expect the ontogeny would reflect somehow that you start out one way and grow in one direction, but that's not the case. The frog hind limb is a completely new pattern." All of the other changes that must accompany a lengthened ankle bone in order for it to be useful—stretching out the nerves and blood vessels and muscles that run along it—happen automatically, thanks again to the rules of development. "The change was instantaneous," Wagner speculates.

As for the tetrapod limb, the way Hox genes behave as they shape it leads Shubin and others to consider it a true innovation in Wagner's sense and not some minor elaboration on a fin. It may have even evolved relatively quickly, without much genetic tinkering. When genes get a signal to produce a protein, they are helped in their efforts by other stretches of DNA, known as enhancers, that may be many thousands of positions away. The DNA strand bends so that the gene and its enhancer come into contact, whereupon the enhancer reacts with the proteins that are copying the gene in a way that speeds up the process. While the humerus, radius, and ulna in an arm are forming, a large group of enhancers work together to help Hox genes do their work. But in the final phase of an arm's development, when the pattern of Hox genes flips and a hand forms, it turns out that only one enhancer seems to assist them. As our lobe-fin ancestors were evolving toward the tetrapod form, this new enhancer may have been recruited into the genetic platoon for limb development. And with that single addition, the pattern of Hox genes might have been abruptly flipped, and a crude sort of hand appeared.

However long it took for the hand to form, though, it lodged itself into the biological landscape. To Wagner, this stubbornness is another sign of an innovation: it changes the rules of the game by becoming the new boundaries inside which natural selection works. Wagner is drawn to the same lingering patterns that Alberch and Shubin were, and Owen before them. Natural selection pares down fingers, it fuses wrist bones, it makes dinosaur shins as tall as lampposts, but it doesn't make triple thighs. After 360 million years of evolution, it can still only vary the fundamental program of development that tetrapod limbs use to grow from buds.

"If you have that much time, why do we still see this structure, this distinctiveness between types?" asks Wagner. "There are changes that are easily reversible, that don't change the rules under which natural selection is happening, and then there are those which lead to new boundary conditions. If everything was reversible there would be no distinct types that are stable over a long period of time. It's a dialectical way of thinking about it. Social revolutions—why do they interest an historian? Because all the political and economic rules under which people led their lives and pursued happiness changed. It doesn't mean that something extraordinary caused them. Most of them were the side effects of slow changes and coincidences, which led to new modes, so that an organizing principle like the feudal production of goods wasn't attainable any more, and another economic force became dominant. I try to think of evolution in a similar way."

Neil Shubin has found one of the most striking demonstrations of how the innovation of the limb imposes a new order. In 1991 a freak freeze turned a pond in Marin County, California, to ice. It killed hundreds of rough-skinned newts and perfectly preserved their corpses. Shubin, who was working at the time at Berkeley, got hold of the newts while they were still frozen, and in the years that followed was able to study 452 of their limbs. In this one gathering of a single species, he discovered an orgy of variation: almost a third of the newts had some dramatic oddity in their limbs. Two or three normally separate wrist bones were fused together in some cases. In others an extra ankle bone appeared out of nowhere. At first it might seem as if variation were running wild and free among these newts, and that working with such rich material, natural selection could create just about anything it pleased. Yet the variation was actually very limited and biased. Fusions and extra bones almost always occurred along the path of the branching limb axis that Shubin and Alberch had recognized in 1986, between connected clusters of mesenchyme cells. And Shubin also discovered that in this frozen pond was a microcosm of all salamanders. One fusion Shubin found in a rough-skinned newt was a standard feature for all of the species in the distantly related family of lungless salamanders. Two of the extra ankle bones can be found in exactly the same position only in 280-million-year-old amphibian fossils.

"That's what you inherit at each step, the rules for building structure," says Shubin, "and the rules are in part encoded in genes and in part coded in the interaction of the things that those genes specify. Sometime you have an innovation that provides a new set of constraints and opportunities. The origin of fingers is no doubt an innovation. You have a new pattern of gene expression that corresponds to a new kind of embryonic development which corresponds with a new set of niches. Vertebrates can now run and fly and do a lot of new stuff. But you're escaping the constraints of history at one level and providing a new set of constraints at another."

Scientists still disagree over what the full picture of developing limbs will look like. Some still argue that a Turing pattern shapes them and that Hox genes only add adjustments to the forming bones by making some mesenchyme cells stickier than others. Others lean toward the idea that the conversation of genes and their proteins draws a map for the Hox genes, which then make mesenchyme cells multiply faster or slower, thereby creating bones of different shapes. But as they run the experiments that will settle the debate, they know something that would have been considered daft fifteen years ago. They are not simply retracing the development of the living hand; they are also seeing how the first hand evolved. Haeckel, dry your tears.

Chapter Four

DARWIN'S SAPLINGS

Whereof Mike Coates was contemplating the many fingers of *Acanthostega* and what they implied for our genes, Jenny Clack was at work on its ears. Her work illuminated a thread of tetrapod evolution just as important as the origin of hands, as well as an ingredient of macroevolution as critical as innovations. But news of the work on ears didn't surge through scientific circles the way the finger reports did. Page through scientific papers about the development of the limb and you can begin to understand why. There you see drawing after drawing, photograph after photograph, of disembodied arms and legs. What happens in the embryo to which the limbs are attached doesn't matter to these experiments. Only what happens inside the limb itself determines its pattern, and that gives these experiments their strength: a biologist doesn't have to keep track of genes at work beyond the shoulder. It is the isolated problem that standard science craves. Clack's work on the ear, on the other hand, was much messier, and the messiness was what made it important.

The best way to understand the ears of early tetrapods is to start with our own. Sound is a gentle shift in the wiggle of molecules, which travels as waves through the air and washes softly up against eardrums. Set against a tiny rod of bone called the malleus, an eardrum can vibrate, and these vibrations pass through a series of bones in the middle ear: malleus, incus, and stapes. The stapes fits in a doorway leading to a maze of tubes filled with liquid and lined with hairlike nerve endings. Some of these tubes (called the vestibular system) maintain our balance by sensing the slosh of the liquid with every tilt and turn of the head. But the stapes makes direct contact with a different, coiled tube called the cochlea. The vibrations that the stapes passes into the cochlea create waves in the liquid that make the hairs sway. The hairs relay the swaying into the brain, where we comprehend it as sound. This kind of ear—common to mammals—is an elaborate model that's good for a wide range of frequencies. Many other tetrapods hear relatively well despite the fact that they lack some middle ear bones; their eardrum connects directly to the stapes. Fishes meanwhile lack a tetrapodlike ear altogether—although they aren't exactly deaf. They have long lines of pits along their head and sides, called lateral lines, that carry pressure-sensitive hairs. They can't hear with their lateral lines, but they can at least sense movements of water produced by other fishes.

Nineteenth-century embryologists took the first steps toward understanding how the tetrapod ear evolved. Searching for homologies between fishes and tetrapods, they found that the stapes in the human ear corresponds to a large bone that helps support fish jaws, known as the hyomandibular. The human jaw opens like a door, swinging on two hinges where it contacts the skull, as muscles anchored to the head and neck pull it back. When a Devonian lobe-finned fish opened its jaw, however, it had to conduct a biomechanical symphony. Its skull was a loose collection of bones held together by ligaments. The hyomandibular served to brace the upper and lower jawbones against the back of the braincase. At the same time it was also helping to flare open the gill flap in order to let the stale water pass out of the head of the animal. As cluttered as this arrangement might seem to us, it worked well for a lobe-fin. Since the muscles and bones it used for breathing and feeding were coupled together, it could hunt prey by opening its gaping mouth and sucking them in.

Yet as essential as the hyomandibular might have once been, it became obsolete in tetrapods. Rummaging around the back of *Acanthostega's* head, Clack found its earliest known incarnation as a small, blobby stapes. Her discovery, combined with older research, shows how the hyomandibular dwindled as early tetrapods changed from the lobe-fin style of eating. The snout grew five times longer—far too long for muscles at the back of the head to bend it efficiently. The braincase fused shut, and the jaw began to make direct contact with the sturdy skull. As biting became simpler for tetrapods, tetrapods no longer needed the hyomandibular bone to support the jaw.

Instead of a big bone linking to the palate, lower jaws, and gill bones, *Acan-*

thostega's hyomandibular-*cum*-stapes had shrunk and become lodged tightly into the back of its skull. It no longer supported the jaw or controlled the gills. Instead, it had become a cranial keystone that held the roof of the mouth to the braincase and formed a piece of the braincase wall itself. Yet the job that it does so well in our own heads—letting us hear—was still in the future, because it was now locked into the skull and couldn't vibrate freely. Only much later did the other bones of the skull become sturdy enough that the stapes could loosen and begin transmitting sounds to the brain.

The tetrapod ear, so unlike anything that came before, is as much of an innovation as the limb. But you cannot trace its history by pretending that it evolved in isolation: the hyomandibular changed into a stapes as part of an interconnected set of transitions. It could evolve as it did only because the tetrapod head was flattening and lengthening and fusing to create a new kind of bite. At the same time, it could change only as the gills were also becoming less important to the animal for breathing. In turn, the shrinking hyomandibular had effects that radiated out to other parts of the body. No longer controlling the gills, the hyomandibular stopped straddling the head and shoulders of tetrapods. The muscles that once connected it to the gill arches could now reattach themselves to the jaw to help open and shut it, and to support its head on its shoulders. In other words, the dwindling hyomandibular let other bones and muscles create the tetrapod neck. The effects of its transformation even reached the limbs. Only when the shoulders were liberated from the head and from the heavy bones that once covered the gills was there enough room for a bigger, more complex shoulder joint, one that made the arms of tetrapods strong enough to walk on.

This choreography of changes caught the attention of Keith Thomson in the 1960s when he did research of his own on how the tetrapod ear came to be. It seemed to him to speak of a common feature of macroevolution, which he named correlated progression. A change in any one part of the tetrapod head and shoulders could not take hold unless natural selection was altering the other parts for other adaptations at the same time. In one sense correlated progression makes macroevolution harder to pull off—how impossibly lucky does a proto-ear have to be in order for all the other necessary changes to all be happening independently at the same time? But correlated progression may actually be able to make macroevolution easier, since the changes in one part of the body can sometimes make other changes more beneficial to an animal. Being able to walk more like a tetrapod could make it easier to catch food with a tetrapod's fused skull, and being able to catch food this way would encourage the continuing evolution of its limbs. This feedback between the evolving parts would make change easier.

Correlated progression can also change an animal's body through the tangled effects of genes. Altering one gene can produce many changes. A mutation to a Hox gene will disrupt the limb, spine, and other parts of the body. If such a mutation creates several changes that together are a boon to an animal, it can speed the pace of evolution. Correlated progression can also happen because of the way different tissues in an embryo interact. Dog breeders depend on this principle for their livelihood: by artificially selecting generations of dogs for different facial features, they have stretched and squeezed the canine head in all sorts of ways, from the upturned pout of a Japanese spaniel to the tapered nose of a Borzoi. In just about every case the rest of the dog's head has successfully changed along with the particular traits that the trainers have focused on. The teeth still lock together, the veins still curve around the jaw (although the dogs can't cope with every change: pugs for example have had their faces so squashed they cannot stop tearing). The extremes that dog heads can reach make clear how their features aren't fixed by some particular genetic map. They respond to any change in surrounding tissue, adapting to produce an organism that can still function as a whole.

Correlated progression, as we'll see, is a common feature of macroevolution. If anatomical structures in an animal are tightly linked together, they change in concert as a lineage passes through a dramatic transition. The ghost of Cuvier lives on in the correlated progression of life: animals are made up not of completely independent parts, but of organs and tissues that work together with intimate integrity. Yet what Cuvier saw as an obvious roadblock for evolution is now a new boulevard of change.

Acanthostega was surpassing all the hopes that Clack and Coates had when they began their study, with almost every bone a revelation. As their preparator, Sarah Finney, slowly set free more and more of its skeleton, they began to turn their attention to the entire beast—not as the sum of remarkable parts but as an actual animal that had to make its way through some Devonian ecosystem. In the air *Acanthostega* was practically deaf, its stapes a mere stub lodged in its skull, but it still had lateral lines that could sense pressure waves underwater as well as any fish. At the same time it was a fully evolved tetrapod in other ways:

it had the four feet complete with an abundance of well-turned toes. This clash of anatomy didn't jive well with Romer's textbook scenario of fish stranded by drought evolving legs as they crawled to the next puddle. The forces driving lobe-fins toward the tetrapod form must have been at work already for some time by the time *Acanthostega* appeared—according to Romer, on creatures forced to spend part of their time moving around on land. Yet even a bullfrog has infinitely better hearing on land than *Acanthostega* did. The clash between fish and tetrapod only became more drastic as Clack and Coates moved from head to tail.

Not far from the stapes, Clack and Coates discovered a strut of bone in *Acanthostega*'s neck, a bit smaller than a matchstick, which they recognized as a gill arch. Amphibians still retain their gill arches, but they serve a new purpose, anchoring the tongue muscles. In cross section an amphibian's gill arch is round, whereas *Acanthostega*'s arch was shaped like a crescent moon. Among living animals only fishes have crescent-shaped gill arches, and for a very specific purpose: they form a groove in which blood vessels can fit, and these vessels bring tired, carbon dioxide–loaded blood to the gills and carry it away loaded with oxygen. In other words, this curve in the bone told Clack and Coates that unlike any other tetrapod, *Acanthostega* still breathed water like a fish.

The longer Clack and Coates looked at *Acanthostega*, the more primitive they realized the animal was. Here was a tetrapod that couldn't be accommo-

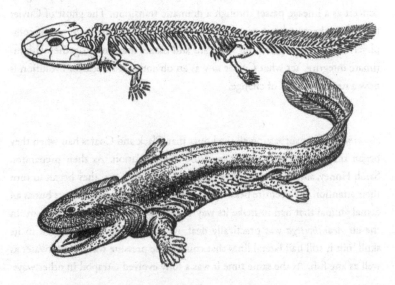

The skeleton of *Acanthostega*, and how it looked in life

dated in Romer's version of evolution. Walking on land seemed close to impossible: *Acanthostega*'s shoulders were eggshell-thin, and all of its weight pressed down on its radius, which was fat and round at the elbow but thin at the wrist. "This is like using a table knife as a pillar, with the blade on the ground," says Coates.

Its ribs were no better—they were stubs that wouldn't have been able to cradle *Acanthostega*'s guts under gravity—and its vertebrae were loose, wide rings around an exposed notochord. Like any self-respecting tetrapod, *Acanthostega* had legs and hips, but its hips were only barely held in place by a single slender rib and some ligaments, rather than being fused to its spine. Its knee and ankle were so flat and stiff that it could have only held its hind legs out to its sides like oars. And while *Ichthyostega* had amazed paleontologists with half a fish's tail, *Acanthostega* had the full version. The long array of rods and fins, a complex mix of two different kinds of bones carefully arranged with muscle attachments, was found not only along the top of the tail but along the bottom. Its tail could drive it forward and brake underwater, but on land it would only be a burden, the delicate tissue along the bottom getting scraped, cracked, and sickly.

Here then was a tetrapod that was as comfortable underwater as any fish but miserable unto death on land. There could be two explanations. One was that tetrapods had evolved the hallmarks of their form while adapting to life on land, and *Acanthostega*'s immediate ancestors had then gone back to the water, reinventing fish gills, a fish spine, and a fish tail. There are certainly many amphibians that have gone back to the water, but they always evolve new *external* gills—feathery blood-rich structures that extend out from their necks. Contrast these with the gills inside a fish's mouth, mounted on arches and accompanied by a flap that flushes out the water that passes over them. To swim, aquatic amphibians can still wriggle their bodies, and their tails sometimes carry a fleshy strip to push more water. But none have managed to reevolve completely the gorgeous intricacies of a fish tail, because evolution can never retrace its steps so exactly. "There's all sorts of jiggery-pokery you have to do to justify saying *Acanthostega* was secondarily aquatic," says Coates. That leaves the other, stranger choice: that the things we associate with surviving on land evolved for life underwater.

Our own tetrapod bodies are such a burden in water—try running on the bottom of a pool and imagine being chased by a lobe-fin as big as a killer whale—that this idea is at first ridiculous. What advantage could possibly nudge some fishes toward a tetrapod body in a place where they already did so well? An answer of sorts was discovered by a crew of Dutch sailors in 1696. On the coast of Australia they encountered fishes about two feet long, each with, they reported, "a sort of arms and legs and even something like hands." The creatures became known as frogfishes because of their reputation for walking on land. In the early 1700s an artist living in Indonesia wrote that he kept a frogfish alive for three days in his house: "It followed me everywhere with great familiarity, much like a little dog."

Frogfishes, which are ray-finned fishes like zebra fish and trout, have nothing to do with the origin of tetrapods, yet their fins have evolved into what Owen would call analogies of legs. In each one, three of the rays have stretched out into shinlike structures, connected at the tip to a fan of shorter, toelike rays. Contrary to old stories, though, these fake legs don't work on land, and a frogfish that finds itself out of water will spread out like a pancake under its own weight. Underwater, where most of a fish's weight is canceled out by its buoyancy, the fins do work well. On coral reefs some species of frogfish gently sink onto their fins, which bend until they take the appearance of feet with toes. The muscles that most ray-finned fishes use to paddle their fins back and forth a frogfish uses instead to swing its fins in a tetrapodlike arc, bending it like a knee and setting it down like a foot. Its pseudo-toes let it grip the coral and lift the other fin to take the next step. In its timing, in the pattern of forces it generates, a frogfish walks like a tetrapod, although it manages to move less than a mile an hour. One species that lives in the vast forests of kelp that grow in the Sargasso Sea uses its fins not so much like a horse as like a spider monkey in a forest canopy: it can actually fold its fan of rays around the stems and swing from frond to frond. In both of these underwater habitats, legs and toes make new ways of survival possible. As a fish swims, its undulations kick up waves that other fish can detect with their lateral lines. A frogfish can walk slowly and imperceptibly and hide by grabbing a rock rather than treading water.

From time to time scientists have pointed out that the frogfish might be an analogy for early tetrapods, but they haven't been taken terribly seriously with-

A frogfish uses its fins to walk underwater.

out fossil evidence to back them up. To Coates and Clack, though, *Acanthostega* could not have existed if evolution had taken any other course. Given the new environment for early tetrapods presented by the likes of Keith Thomson, such an underwater origin makes sense. On the edges of continents plants were beginning to form the first forests on earth, and the coastal waters began to thicken here and there with swamps. Trees wept their leaves into the water and dropped their boughs. Their roots stopped up sediment and created velvety soil. Invertebrates burrowed into the muck to feed, and fish followed. Lobe-fins did well in this habitat; some grew longer than pool tables. Others skulked on the bottom, trying to make their way over and through the muddy wreckage. Thanks to the way their genes built their fins, some lobe-fins were able to ornament their fins with skeletal projections—fingers and toes—which helped them lumber on the swamp floors. After some millions of years, one stock gave rise to *Acanthostega*. For the most part it lived underwater, content to breathe most of the time through its gills, climbing over logs or clutching a rock while waiting for an approaching fish. Once the prey came into the view of *Acanthostega's* overhead eyes, it released its anchor, used its powerful tail to race upward, and swung open its long mouth. In particularly shallow water, it could push up on its front legs to lift its head up to the surface and gulp air.

Before the discovery of *Acanthostega*, many paleontologists had assumed that the evolution of the tetrapod body and the emergence from water were one and the same. But this fossil splits the two apart. Tetrapods already had most of the basic evolutionary innovations that they would need to live on land dozens of millions of years before they ever made land even a brief part of their lives. What looked like adaptations have become exaptations of the most dramatic sort.

The haul of rocks that Jenny Clack had brought home from Greenland included not only tetrapods but fishes as well, and her old student and field hand Per Ahlberg took them to his new job at Oxford University in 1990. For a few years he had been studying a devious little group of lungfish relatives called porolepiforms, and he was still sorting out their relationships. Before traveling to Greenland, he had gotten many of his fossils from drawers of museums, where they had lain unidentified for decades. When Ahlberg came to Oxford he opened their drawers as well, looking over anonymous fossils that had been dug up along the Longmorn Burn in Scotland at a spot called Scat Craig. The bones dated back 370 million years, 7 million years older than *Acanthostega* and *Ichthyostega*.

"The locality was discovered back in the 1820s," Ahlberg says. "There were twenty years of intense collecting by gentleman amateurs. This was about the time that people realized there were fossil fishes in Scotland, or even fossil fishes, full stop. There were amateurs all over the place in Scotland, and at Scat Craig they were finding isolated bones, but in good condition. A lot were collected and papers were published, and then these collections ended up in different places. There's a small collection at Oxford which I found still uncataloged after 130 years. It was collected by the master of Worcester College and his daughter on vacation, and had been presented to the director of the University Museum with a nice little letter asking if his rheumatism had gotten better."

As Ahlberg looked through the master's collection, he quickly found a good piece of a porolepiform fish and settled down with the Scat Craig material to look for more. "Once I started going through specimen by specimen, I had to stop and think about each piece," he says. "And I was struck by a piece of snout which was flat." It was a detail that most paleontologists wouldn't have noticed, and the only reason that Ahlberg gave it any thought was that a few years earlier he had been crouching on a bare mountain in Greenland, trying to decide whether the bones he was finding were Devonian fishes or tetrapods. He was now sensitive to obscure things like 370-million-year-old snouts. "If you only had experience with later tetrapods and you looked at the Scat Craig material, you'd think, Oh—fish."

The shape of this snout, Ahlberg realized, meant that it actually belonged to a tetrapod—the oldest tetrapod bone known. Looking in the drawer he found more tetrapod bones, and soon he set out on a pilgrimage to search the diaspora of the Scat Craig fossils in museums throughout England and Scotland. His collection grew year after year, jaws and snouts and tibias and bits of thigh, all hidden in cabinets for 130 years. He has now tentatively reconstructed the animal (which he calls *Elginerpeton*, after Elgin, a town near Scat Craig) as a five-foot-long pointy-headed beast, with legs twisted into oars even more drastically than *Acanthostega*'s.

In the 1970s the only Devonian tetrapod known with any certainty was *Ichthyostega*. Now the tetrapod zoo has abruptly filled with a number of animals such as *Elginerpeton*. Rather than lonely windows into life in the Devonian, the fossils can now actually show us the origin of tetrapods step by step. But in order to watch this flip-book movie, paleontologists have had to figure out exactly how the animals are related to one another. It is time to do some taxonomy.

Taxonomy can sometimes seem rather quaint, with the texture of old glass-fronted cabinets in which an eighteenth-century naturalist might set his seashells. Taxonomists debate in whispers about whether a shrub represents a new species, a new tribe, a new infraorder, a new subclass. They have a strict protocol for how to strike out redundant names given to fossil tails and shells that they later realize belong to the same turtle. Does that extra tooth in that vole's jaw mean we're dealing with a new species, or is that just a local variation? In the reconstruction of life's history, taxonomy might seem about as crucial as taxidermy. In fact it is the skeleton on which the body of evolutionary biology hangs. Without first knowing exactly how organisms are related to one another, it's impossible to understand how new body plans came to be.

Taxonomy has also seen some of the most furious clashes in the recent history of biology. Scientists like Linnaeus grouped together species by their similarities, into genus, order, family, and higher ranks, but nowhere in their system could they take into account the fact that organisms are similar to one another either because they descend from a common ancestor, or because they are unrelated but converged on the same kind of body. The class of mammals is made up of orders such as primates and rodents, but which orders are more closely related to each other, and which are the most primitive? After it became clear that the patterns of taxonomy are evidence of evolution, biologists tried to reconcile Linnaeus and Darwin by looking for similarities between different animals, living and fossilized, and sketching out a possible transformation from ancestors to descendants. Once done, they would sketch out their trees of life. Until the 1970s this was essentially how everyone used taxonomy to understand evolution. Gradually, though, word spread of a German entomologist named Willi Hennig. He had written an obscure book about classifying insects that was translated into English in the 1960s; short on clarity and long on words of his own making, the book offered a new way to do taxonomy, and one that could be explicitly tested.

Imagine you've just come across three butterflies, wings opening and closing gently, on a porch screen. One has short antennae, a stunningly purple body, and a webbed pattern on its wings. The second has long antennae, an equally stunning purple body, and a giant spot on the bottom of each wing. The third has the same purple body, long antennae, and red specks on the edges of its wings. How are these butterflies alike and not alike? They all have a pair of wings and stunningly purple bodies. They each have a unique pattern on their wings. But the two long-antennae species are a match, to the exclusion of the third. You can show this likeness as Hennig would, with a tree. Think of it for now as nothing more than a visual representation of logic, like a Venn diagram with its overlapping circles.

Since all of these butterflies are a product of nature—and thus of evolution—they must all be descended from a common ancestor. That ancestral butterfly species split into descendant species, and they in turn split into still more. New traits were evolved, some old ones were lost. Many species became extinct, and of the living, only three happened to show up on your porch screen. Unless all three butterflies had evolved at the same time from an ancestor—a highly unlikely event—two of them must be more closely related to one another than the third. And that, declared Hennig, means that this logical tree also works as an evolutionary one. It shows how the butterfly lineages may have branched apart and evolved their distinctive traits, such as long antennae and wing patterns.

Of course, evolution does play tricks. Lineages can evolve a trait and then lose it. Animals that are only distantly related sometimes evolve into surprisingly similar forms. They may be trying to make the same sort of living in the same sort of environment, or the kind of genetic variability that they have to work with may be limited and tends to produce the same kinds of anatomy. Perhaps the two butterflies with long antennae didn't descend from an ancestor that had them, but each grew their antennae long independently. To overcome these sleights of biological hand, scientists rely on what's known as the principle of parsimony. When they want to study how some organisms evolved, they start with the simplest hypothesis that's consistent with the known facts. Simplicity in this case means finding a tree that calls for the fewest evolutionary steps—

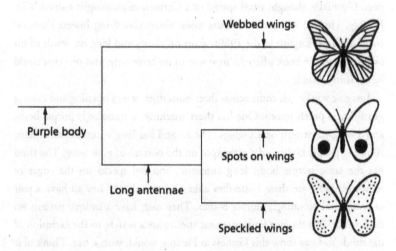

A cladogram of three butterflies

either acquiring a new feature or losing an old one—while still grouping together species by certain traits. In a case such as the three porch butterflies, with so few species and features to study, it's hard to get a good idea of what their true tree is. But when a scientist studies more features on more species—particularly by comparing the group in question (such as butterflies) to species outside the group (say, beetles and grasshoppers), the signal of the true relationships often becomes clearer, swamping the noise of convergence.

Hennig's evolutionary tree is a pretty abstract piece of timber. On its own it can't link a species to a particular ancestor. The webbed-wing butterfly has a deeper root than the other two, but that doesn't mean it's more primitive than they are. On these kinds of trees the common ancestor of two lineages appears as nothing more than the segment leading up to the node where they join. The common ancestor of the three butterflies was probably a butterfly with short antennae and a purple body—we can say no more. Nor does such a tree contain all the branches of other species that descended from the same common ancestors. Yet in order to explain the relationship of the three butterflies on the porch screen and how evolution produced it, this is enough. And by extension, it is enough to set up a new classification for life.

An early taxonomist would classify elephants in their own order because they all have unique features such as trunks and tusks. One of Hennig's trees would have shown how trunks and tusks must have arisen in some long-dead population of proto-elephants and were inherited by their descendants. We class elephants in a larger group that also includes animals such as otters and moles and humans because they all possess hair, a warm-blooded metabolism, mammary glands, and other traits: they are all mammals. No other animal can offer these traits in such a combination, and they demonstrate that mammals all descend from a common ancestor. Among any group of species, we can find traits that some share and others do not, and other more primitive traits that they all share. In organizing animals by evolutionary descent this way, Hennig pulled biologists away from the Linnaean system and toward branches of life, called clades. His evolutionary trees he called cladograms.

A growing number of biologists came to find Hennig's method, now called cladistics, to be a more explicit way of studying evolution than previous ones. Cladists have no choice but to lay out all of the assumptions they make about how species are related, trait by trait, and in a way that can be tested. But even those who admired cladistics didn't relish building cladograms in the early days. They had to construct them with pencil and paper, adding up the steps that produced their data most parsimoniously, and the fact is that cladistics has a nasty habit of exploding. If you examine three butterflies, you need to compare three different possible trees, but if you analyze a dozen butterflies, millions of possible trees exist, and with a hundred species (only a pinch of the twenty thousand butterfly species in the world) you are faced with more trees than there are atoms in the universe. But just as Turing's ideas of how patterns can form without maps have invigorated embryology, mathematics has been a huge help to the new taxonomy. Sorting through trees is just the kind of simple, eye-glazing chore that computers do well. A biologist can load a cladistics program with information on dozens of species, inspecting a hundred or more parts of their anatomy and noting whether each has a certain spot, bump, or furrow. The computer does not actually look at the entire cosmic forest of possible trees but uses statistical shortcuts to find a few million of the best candidates, and then hunts for the simplest tree among them.

Some paleontologists quickly took up cladistics, but many were suspicious. Cladists showed up in the museums speaking an abstruse jargon, carrying their rattling satchels full of new words such as *apomorphies, synapomorphies, autapomorphies,* and *symplesiomorphies.* They were so taken with their method that they seemed like missionaries for the Church of Hennig, wanting to throw out long traditions of taxonomy despite the fact that the conclusions that came rolling out of their computers often only confirmed what paleontologists had known for decades. Some cladists even claimed (wrongly) that fossils were irrelevant to building cladograms because they offered so few traits to analyze, compared with living animals with all their details of flesh and blood.

Cladists in turn saw many of their paleontological opponents as lost in ancestor worship, spending their time trying to find exactly which species begat which species, or which class arose from which class. And yet the information that we have in fossils and living animals cannot, logically, tell us whether a species is an ancestor of another species or a merely a close branch. Even if you find two nearly identical fossil fish species in adjacent layers of rock in the same hillside, you can't say with certainty that the older is the ancestor of the younger. The cladists argued that, rather than chase these dreams, it was more important to find the most objective, testable way to organize life.

Cladistics is now a standard tool in paleontology, although down museum halls you can still hear grumbles about its sterility. While cladistics has indeed simply confirmed some long-established ideas, paleontologists have also used it to argue for some controversial relationships—supporting the view, for in-

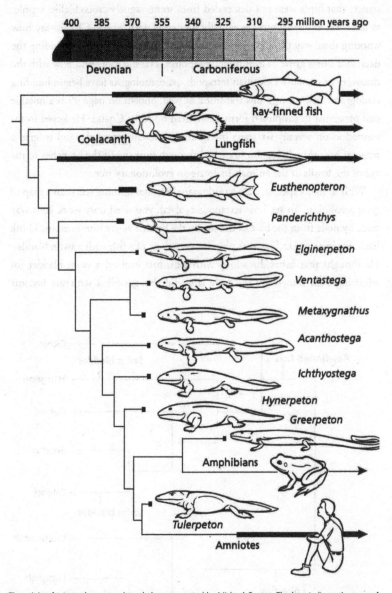

The origin of tetrapods, as seen in a cladogram created by Michael Coates. The bars indicate the span of each lineage's known fossils, and the thin lines mark the branchings that have not yet been documented. Some of the animals (such as *Hynerpeton*) have scant fossil records and are illustrated to resemble their close relatives.

stance, that birds were not descended from some vaguely crocodilelike reptile, as once was thought, but are actually dinosaurs with wings. Cladists are now working their way through the animal kingdom, their computers grinding the data, and a new grove of evolutionary saplings is taking root. And now with the discovery of so many Devonian tetrapods, paleontologists have begun building cladograms that capture this transition as well. Shown on page 99 is a notable one presented in simplified form, computed by Mike Coates. He keyed in information on seventy-six traits in eighteen different tetrapods, and to give a sense of how this cladogram branched through time, he fit the branches to the age of the fossils of the animals to create an evolutionary tree.

With so many species, such a cladogram becomes nothing less than a map of macroevolution. To see how tetrapods evolved, you need only work your way node by node from the base of the tree to the branch you're interested in. Think back, for instance, to Darwin's pleasant genealogy of a fish with a swim bladder. He thought that fishes, breathing with gills, first evolved a swim bladder for adjusting their buoyancy, and only later did this gas-filled structure become

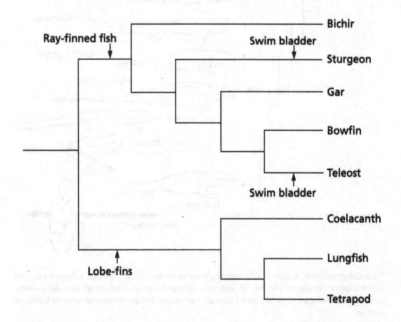

Except for sturgeons and teleosts, all major lineages of ray-finned fish and lobe-fins have lungs instead of swim bladders.

lungs in lobe-fins. For over a century most biologists agreed with him, but a cladogram of the major division of fish that have lungs or swim bladders shows that this belief is a trick of the taxonomic eye. The tree shown on page 100 shows the major branches of fishes that have swim bladders or lungs (coelacanths do not have lungs, but do have chambers, now filled with fat, which seem to have once been lungs). Each branch of this tree represents a distinctive lineage that reaches back hundreds of millions of years, although some of them contain few living members. Bichirs, for example, are limited to a few species in the swamps of West Africa. By contrast, teleosts are the most diverse group of ray-finned fishes, including most of the species we're most familiar with, from goldfish to salmon.

The most parsimonious explanation is that the common ancestor of lobe-fins and ray-finned fishes had lungs, and that among a few lineages the lungs turned into swim bladders. It just so happens that today the lineages that can't breathe air are the most common. Darwin and the rest fell victim to hidden assumptions—that lungs must be more sophisticated than swim bladders because we lofty humans have them, or that because teleosts are so common they must be typical, and therefore primitive.

The old bladder-to-lung transformation made sense when people thought that the forerunners of tetrapods lived in drought-prone ponds that could become stripped of oxygen or disappear altogether. Where gills couldn't work, lungs could. But now Barrell's parched landscapes have washed away, researchers such as Keith Thomson have shown that early lobe-fins lived in the ocean or in brackish inlets, and others have discovered that even earlier ancestors of ray-finned fish and lobe-fins—among whom lungs first evolved—lived farther out at sea. In the open ocean the waters are stirred so continuously that a typical fish never has to worry about running out of oxygen.

A clue to the true evolution of lungs may lie in the fact that you can kill a trout by making it swim hard. Lungless fishes such as trout pump their blood in a simple loop, sending it from the heart to the gills, where it fills with oxygen, and then on through the rest of the body, nourishing the swimming muscles. By the time the blood gets back to the heart, most of the oxygen has been used up and this saintly muscle has to make do with what's left over before pumping the blood back to the gills for a refill. As a trout swims faster, things only get worse: its swimming muscles hunger for more oxygen, leaving less for the heart which is working even harder than before. After a few minutes of racing, most fish without lungs die.

On the other hand, as Colleen Farmer of Brown University has pointed out recently, fishes with lungs have more stamina. Their blood pumps through two circulatory loops. The gill loop is much the same as in a lungless species (from gills to body to heart to gills), but when it breathes air through its lungs, the blood travels from these lungs to the heart first rather than last. Thus when a fish such as a bowfin has to swim long distances, it takes breaths of air from time to time to keep its heart from becoming exhausted. Perhaps, Farmer suggests, the ancestor of lobe-fins and ray-finned fishes was a powerful, fast-swimming predator that would wreck its heart if it had to depend only on a simple gill-body-heart-gill loop. It developed lungs as pockets in its digestive canal that became dense with blood vessels. Because they could absorb oxygen from the air into their lungs when they surfaced and keep their heart nourished, these fishes could swim harder and longer than other animals, and presumably catch them.

According to this scenario, by the time tetrapods first evolved, their lungs—which seem so perfectly suited for life on land—had existed for 60 million years. If Farmer is right, the question about lungs that really needs answering is, why are teleosts—which lost their lungs—so much more common than fishes with them? Breathing requires a fish to come to the surface to put its mouth to the air, and until 220 million years ago this was a safe proposition. But after that time, the sky began to fill with predators—first scaly-winged pterosaurs, and later birds—that snatched fish that they could spot while flying over water. It's possible that teleosts sank underwater for good, their lungs turning to swim bladders, and thrived.

Once lungs came into existence, breathing itself evolved through stages as lobe-fins changed into tetrapods. Cladistics allows researchers to reconstruct a history of ventilation by studying living animals. A lobe-fin—lungfish, for example—breathes with a kind of two-stroke mouth pump. To pull in air, it expands its gill arches so that its mouth expands. The pressure of the air in its mouth drops with the increase in its volume, sucking in more air from both mouth openings—stale air rising from the lungs and fresh air coming in through its nostrils. Some of this mixed air escapes through its nose, and then the fish squeezes its mouth cavity shut to force the remaining mix of new and old air back into the lungs.

Amphibians still use this pump, although they've added some new touches. As an amphibian widens its mouth cavity to draw in air through its nose and lungs, it also tightens muscles running along its flanks. This compresses its lungs, forcing the air out more quickly. Then, like lungfishes, the amphibian squeezes its mouth shut to swallow air, but it can swallow the air with added power thanks to large holes in the roof of its mouth that are filled with muscle-bound membranes. By contracting these muscles, the amphibian can press down on the air in its mouth and push it down harder into its lungs. (Some frogs squeeze so hard that their eyes—which sit on top of these mouth-roof membranes—disappear from view each time they inhale.)

The other branch of living tetrapods—today's mammals, turtles, birds, and various reptiles (collectively known as amniotes)—have built their own breathing system on the lungfish plan. An iguana is probably the best living analog to how primitive amniotes breathed. Having lost the gill arches of early lobe-fins, it can no longer use its mouth as a pump. Like amphibians, it exhales by squeezing its lungs shut with muscles running along its flanks, but unlike amphibians,

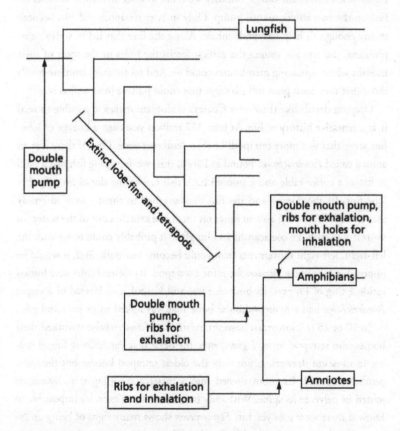

It's possible to reconstruct the evolution of tetrapod breathing on a cladogram.

amniotes have full rib cages lined with another set of muscles. After exhaling, they can expand the rib cage, creating low pressure in the lungs to breathe in. (Once early amniotes had shifted their breathing system from their mouths to their rib cages, different lineages invented variations on the theme. We mammals, for example, still use muscles to open our rib cages, but we also evolved a diaphragm that pulls in air more efficiently.)

These clues from studies on living animals are a test for the way paleontologists arrange their cladograms. Evidence from the fossils of extinct animals along these diverging forks agrees well with the data from living animals. Tetrapodlike animals such as *Eusthenopteron* still had ribs that barely poked out from their spine and obviously couldn't have helped in breathing, suggesting that they took in air like lungfish. Ribs grew larger on the first tetrapods such as *Elginerpeton* and *Acanthostega,* but they were still so small that these animals relied on the two-stroke mouth pump. Only in later tetrapods did ribs become sturdy enough to help tetrapods inhale. Along the line that led to today's amphibians, you can see among the earliest fossils the holes in the roofs of their mouths where squeezing membranes could fit. And meanwhile, amniote fossils show that they soon grew full rib cages that could pull air in as well as out.

Draping details like these over Coates's cladogram makes it possible to read it as a tentative history of life. At least 377 million years ago a lineage of lobefins arose that was more tetrapodlike than *Eusthenopteron.* One of these was an animal called *Panderichthys.* Found in Latvia, this two-foot-long fish had a skull as flat as a coffee table and a smooth back that lacked the dorsal fins of other lobe-fins. Its shoulders—and the fins that attached to them—were so sturdy that it might have been able to move on them like crutches out of the water for short distances. Like coelacanths and lungfish, it probably could move with the left-right, left-right movements that would become our walk. Still, it would be impossible to mistake *Panderichthys* for a tetrapod. Its toeless limbs were buried inside a ring of fin rays, its braincase was still hinged, and instead of a stapes *Panderichthys* had a hyomandibular bone that was linked to its jaws and gills.

In 10 or 15 million years, however, relatives of *Panderichthys* reworked their bodies into tetrapod form. *Elginerpeton,* the beast that Per Ahlberg found hiding in museum drawers, is not only the oldest tetrapod known but the most primitive as well. Its snout turned into a massive snapping trap, ligaments joined its pelvis to its spine. With only fragments of its limbs, it's impossible to know if there were toes yet, but *Elginerpeton* shows many signs of being an intermediate between lobe-fins like *Panderichthys* and later tetrapods. Its rear legs were twisted so much that its knees (if it had them) would have pointed to the ground, making the legs useless for walking but good for rowing. Judging from the fact that *Elginerpeton* was five feet long and hunted on river bottoms, one can assume that the first tetrapods must have been moderately successful at living like a lobe-fin. Within a few million years *Elginerpeton* was gone, but new kinds of tetrapods were evolving all around the world. One branch included an animal called *Ventastega,* known from parts of its skull and a few other fragments found in Latvia, as well as the even more obscure *Metaxygnathus* from Australia. At the time these two landmasses were thousands of miles apart along the equator, indicating that tetrapods were moving quickly along the tropical coasts, invading estuaries and rivers as they went. Yet despite spreading around the planet, this lineage of tetrapods soon died out.

Closer to our own heritage were two tetrapods that appeared in Greenland, which was also near the equator: *Acanthostega* and *Ichthyostega. Acanthostega* had fingers and toes on its limbs, and its braincase had sealed shut. It had a rigid, long snout that no longer needed a hyomandibular bone for support, and this bone had been reduced to a stapes locked into its skull. It's possible that *Elginerpeton* had already reached these landmarks, but there's too little of its skeleton to tell for sure. In either case, the fossils speak to a flurry of change, one that— as the work of Neil Shubin and others shows—was based on the vagaries of genes and development. The lobe-fins like *Eusthenopteron* and *Panderichthys* used Hox genes to pattern their spine and fins. In the first tetrapods, mutations flipped the Hox pattern in the limb and expressed it along the far edge of the fin, producing a new set of segmented rods.

Ichthyostega, once seen as a missing link between fish and tetrapods, now seems to be another weird experiment in limbs. In some respects it was more like later tetrapods—its fishy tail was already half gone, the bones of its forearms were of equal length so that it could walk on them, it had large ribs, and its hips had deep sockets to hold the balls of the femurs. Yet according to one of Jarvik's Swedish colleagues, Hans Bjerring, as well as Coates, its hind legs were thrown back behind its body like a seal's. Although it had sturdy ribs like later tetrapods, it seems to have taken them to extremes, rendering them wide slats. It's possible that it used them as a manatee uses it own oversized ribs, to counteract the buoyancy of its lungs and stay underwater.

In a sense, both *Acanthostega* and *Ichthyostega* were living fossils in their own day. *Hynerpeton*—the Pennsylvania tetrapod that Ted Daeschler discovered in its own lush bayou—lived a few million years *before* them, and yet it is closer to tetrapods alive today than they are. It had evolved even more powerful shoulders than *Ichthyostega*, although with only a single shoulder bone to *Hynerpeton's* name, it's an open question what its hind legs looked like. Perhaps it could only drag them like *Ichthyostega*, or perhaps it held them more upright. In any case, tetrapods may have given up breathing with gills by the time *Hynerpeton* came into existence, as evidenced by its shoulder, which shows no signs of having been connected to a gill chamber. It's possible these tetrapods could gulp air and hold it underwater, or perhaps they were living in such shallow muck that they could often move through the water with their noses raised high. It's possible that *Hynerpeton* was already using its ribs to help its inhalation, because now only its lungs could bring it oxygen. The change in its gills also changed the way it ate. Earlier tetrapods could have sucked in their prey by pushing water through their gill slits. But once their gill chambers had disappeared, they could only lunge for their meal.

Around the same time as *Hynerpeton* there were probably a number of other tetrapod species splashing around that we have not yet discovered, but only one of its close relatives gave rise to all the tetrapods that live on earth today. Coates believes that by 355 million years ago this progenitor's descendants had split into two major branches, one that would lead ultimately to amphibians such as frogs and salamanders, the other to amniotes—birds, mammals, and reptiles. Both groups, however, began very much underwater. When we think of the journey of vertebrates onto land, this new phylogeny forces us to imagine two separate odysseys. Coates puts such an early date on the split because of the Russian animal known as *Tulerpeton*. Coates has worked with the animal's discoverer, Oleg Lebedev at the Paleontological Institute of the Russian Academy of Sciences, and they now suggest that this Devonian six-fingered animal is actually an early relative of living amniotes, despite the fact that its fossils were found in rock formed in shallow ocean waters. At some point after this split, amphibians began to evolve their powerful mouth-roof muscles for breathing, while amniotes began to depend more on their ribs.

As the tetrapods evolved, the role of Hox genes in helping shape their bodies changed by hiccups. Lobe-fins had backbones that were organized into far fewer distinct regions than tetrapods. Hox genes probably helped set their boundaries, as well as determining where their fins should go and helping to build them as an axis with branches. In the first tetrapods the Hox genes began also to switch on in a new place—at the far tip of the developing fin—and there they flipped the pattern in which they produced their proteins, creating fingers. But there's not much difference between an *Acanthostega* thumb and little finger—just as there isn't much difference between a vertebra in its neck and one in its midsection. Genes such as Hox hadn't yet carved out distinct identities in either part of the tetrapod body.

The Hox code became more detailed in its work on amphibians and, independently, even more so on amniotes. Preliminary evidence suggests that amphibian hands fluttered between four and five fingers before most lineages settled down to four. Among the first amniotes there was just as much digital uncertainty (consider *Tulerpeton's* six fingers), but apparently by 325 million years ago amniotes settled on no more than five. Tetrapods may have become more standard-issue in their number of digits as Hox genes began crafting each into a distinctive form. Certainly the same care was evident in the spine. As tetrapods moved from a swimming body to a walking one, they needed a more complicated back: the neck had to be flexible and provide anchors for muscles that moved the head; farther toward the tail the vertebrae had to support a powerful rib cage; farther back still, the vertebrae needed to fuse to the hips. Hox genes set the identities of all of these regions.

Whenever paleontologists find a group of fossils that seem to capture a giant transition like this one, they fret that they may be missing the full story, that the animals they know actually belong on remote branches. But that doesn't seem to be the case with the first tetrapods. "The evidence we have suggests that we're seeing the transition in more or less real time—this is the pace at which things are happening," says Per Ahlberg. "It doesn't look like tetrapods are evolving much earlier somewhere else in the world. The emergence of tetrapods is accompanied by lot of experimentation in different directions. And then by the end of the Devonian there's a certain winnowing out. If you had probed people in the seventies they would have admitted they assumed *Ichthyostega* was typical. But now we see all sorts of peculiar things coming out."

Our ancestors have made it to the point where they look like us in the basics: they breathe air without gulping, they can walk, they have a neck. Yet they still linger in the shallow swamps, perhaps slithering in the night to a new stream nearby, with dry land still the home of millipedes and earthworms. We are left still asking how our exapted ancestors came ashore, and how we came to thrive here.

The cladogram of Mike Coates suggests that tetrapods split into two lineages in the ocean and rivers and came separately onto land—the lineage that would produce us amniotes and the one that would produce the amphibians. The first amphibians surged into more than a dozen major lineages by 350 million years ago. In some cases they looked like yard-long tadpoles with webbed feet as adults. Others grew longer than a canoe, and others lost their legs as they evolved hundreds of vertebrae on their snaky spines. By 300 million years ago, some of them were spending a fair amount of their time on dry land. The evidence is in their bones: some had scales, they stood on strong legs, and they had pared down the stapes to a rod that could conduct airborne sound into their ears. Yet by 210 million years ago these ancient ones were gone. In the fossil record, the living branches of amphibians sprang up soon afterward, looking much as they do today—the Jurassic frogs leaped, the salamanders crawled, the legless caecilians burrowed.

Living amphibians offer the only information beyond bone for how their ancestors lived. Most live in the damp meniscus of ponds, forest floors, and rivers, but a few species are scattered along a wider spectrum. A few toads swim in the ocean and a few others survive in the desert. Salamanders called axolotls live in mountain lakes where they never need to take a breath of air, the legless caecilians live underground like carnivorous earthworms. What makes these amphibians most distinctive from amniotes is their egg, which is more like that of a fish. A growing amphibian embryo lies inside a porous membrane that in turn is coated in a thick rind of jelly. Oxygen and water pass through the jelly to sustain the embryo, and wastes such as carbon dioxide can pass out. Amphibians lay batches of eggs that range in number from thousands down to just one, usually as a globby mass in freshwater. There are plenty of exceptions, though: some eggs are buried in the skin of their mother's back, while in other cases she swallows them and they grow in her stomach. Some grow inside their father's throat or as they cling to his thighs. The young of the Alpine salamander grow inside their mother's uterus, feeding on secretions oozing from its walls, and emerge in adult form—after a pregnancy of three years.

The design of these eggs puts constraints on amphibian evolution. If they are not laid in the right environment, they can dry out, and if they are any larger than half an inch across, the thick blanket of jelly begins to smother them and can no longer keep their structural integrity. Many amphibians are therefore born as small larvae that have to live underwater until they can metamorphose into adults. A few amphibians can emerge from their eggs as miniature adults, but they are at the edge of amphibian performance—as embryos they have to use oversized gills or tails loaded with capillaries in order to pull in as much oxygen as possible.

The forerunners of amniotes looked pretty much like the amphibianlike tetrapods, some so similar that the only way to tell them apart is to compare their neck bones or their palates. By 340 million years ago, the amniote ancestors were spending time on land, and within several million years the first official amniote was born—the last common ancestor of all amniotes alive today. There are three great lineages of amniotes that have survived to the present day: the synapsids (represented by mammals such as ourselves), the parareptiles (turtles), and the true reptiles (birds, crocodiles, snakes, and lizards). It's a lot easier to think of the things that a crow, a human, and a Galapagos tortoise *don't* have in common than the ones they do, but there are a fair number of traits we all inherited from our common ancestor. We all have tough skin, intestines that can absorb water, big lungs, and tears. Most striking is the way we are born. Rather than being encased in jelly, an amniote embryo grows swaddled in membranes stuffed within membranes. Tucked into these sacs are a yolk and three different tissues. One gathers up the urea and other wastes of the embryo, another pulls gases into the egg and pumps them out, and a third bathes the embryo in fluid (this last one is called the amnion; it gives amniotes their name). The primordial amniotic egg was probably like the rubbery-shelled ones laid by turtles. Some mammals evolved a system so that they carried the embryo and membranes inside the womb, while reptiles made their shells a hard, calcium-rich layer. Even in its most primitive form, though, an amniote egg allows an embryo to grow far larger than an amphibian's. Supported by a shell and kept fed and clean by its membranes, it can develop into an air-breathing miniature version of its future adult self instead of a tadpole.

With these facts about living tetrapods in mind, we can travel back 340 million years again and ask, why did tetrapods come on land? Food might seem like the obvious lure, but it was a distant one. The early tetrapods were all big, well-designed fish-eaters. Rippling their tail fins or writhing their bodies, they could shoot through the water and gulp prey, snatching them in broad, flat-topped mouths ringed with rows of teeth and long fangs. Forests were stretching across more and more of the continents, changing sunlight to wood and root, but these underwater hunters didn't have the means to feed on them. As for insects, a tetrapod barely able to carry its sizable body—ranging three to six feet long—across land wouldn't have been much of a threat to a skittering silverfish, nor would such a big animal be likely to calm its hunger with such a little meal.

Paleontologists have wondered for a long time if it was not a tetrapod that reached land first but its eggs. Alfred Romer was of this opinion, and he offered one of his well-turned tales in 1957. The early amniotes, living in the water, laid eggs much like those of amphibians, and so the eggs were subject to the same evolutionary pressures that crafted legs and lungs. He argued that these adaptations evolved to let a tetrapod survive a drought by breathing air and crawling to a surviving pond. Romer assumed that these early tetrapods laid eggs as many frogs do today: a female laid them in freshwater, a male sprayed sperm on them, they developed in jellied blobs underwater, and they emerged as larvae with gills. Droughts would be fatal to them, since exposed eggs would shrivel and die, and if the aquatic larvae hatched on dry land they would be helplessly stranded.

What amniotes managed to do, Romer decided, was evolve an egg that could survive a drought. Its shell and its interior membranes resisted drying out, and from it hatched a baby that was already prepared to breathe air, even though it probably settled promptly into a life in water. Once again, Romer saw a way that coping with an immediate crisis led to a creation that would ultimately let animals live in a very different way: only much later did the adults follow the eggs on shore, and amniotes subsequently spread far across the land. "Today," he announced, "a variety of amphibians are struggling (so to speak) to attain some type of development comparable to that which the reptile ancestors achieved eons ago, but their efforts are too little and too late."

Like some of Romer's other ideas about early tetrapods, this one sags on a bad foundation—namely, the alleged Devonian droughts. In their tropical homes, the ancestors of amniotes didn't face certain death by desiccation every year, nor did their eggs. In fact, some basic physics shows that early tetrapods could have come ashore to lay eggs without an amniotic egg. Several living amphibians bury their eggs a few inches underground, with a peace of mind that comes from the way water behaves in dirt. The flow of water across the boundary of a buried egg depends on many factors—the chemistry of the egg and the soil, the dampness of the soil, and the ease by which water can flow through the soil pores. It turns out that water won't leave an egg unless it is buried in extremely dry soil, which was an unlikely danger for Devonian tetrapods living in a coastal lagoon. Soil would have offered some advantages to an early amphibian looking to lay eggs. Soil keeps an egg supplied with reliable levels of water and oxygen, and its temperature is almost constant day and night. In a Devonian landscape devoid of vertebrates this would have been a much safer place to lay eggs than in water where a fish could eat them.

There was thus ample incentive for an early tetrapod to come on shore, if only to bury its eggs and slide back into water. The larvae that hatched from its eggs would be carried back as well by floods, or perhaps they could wriggle their way to a stream. To this extent Romer still seems to have been right. But the first eggs laid on land might not have been amniotic ones; they may have been the old-fashioned amphibious type. It's even possible that the earliest tetrapods evolved digits not only to move underwater, but to dig burrows on land.

Somehow, though, the amniotes did evolve, complete with their amniotic egg, and they have done very well for themselves by almost any measure. For over 280 million years they have been at the top of the ecosystem on land. No bloodthirsty newt chews up ocelots or javelinas. And although there are up to 3,000 species of amphibians alive today, there are 18,500 amniote species. Was the amniotic egg the secret to their success?

Success is a tricky thing to define in biology. Throughout the fossil record, you can find cases in which a lineage of animals evolves something never seen before and promptly spreads out and diversifies at breakneck speed. Among living animals, the most diverse clades often share some trait that seems like it ought to make them successful. Lizards such as gekkos are diverse, and they also are notable for the pads they evolved on their toes that let them stick to vertical surfaces. Perhaps gekkos became diverse because they could then live in trees, safe from predators, while padless lizards would tumble to the ground. One order of birds, the passerines, far outnumbers other orders today, and they are notable because of their feet, which have ligaments that make them spring automatically into a perch. Perhaps, some people argue, that is why they thrive in trees alongside the gekkos. Plants that ooze resin and other gummy fluids, which entomb or poison hungry insects, are more diverse than species that can't. It's hard to find a patch of land without some kind of rodent (they come in two thousand species), whereas multituberculates, a group of rodentlike mammals, have been extinct for 30 million years. Rodents won out, the argument goes, because they evolved continually growing teeth, which let them gnaw at hard seeds and nuts without being crippled by permanent chips and cracks.

Biologists often refer to these secrets to success as "key innovations." While they seem to be an important piece of macroevolution, they call for a large serving of skepticism because correlation doesn't equal causation. Researchers have offered cases like rodent teeth, bird's feet, and gekko pads without putting them to any kind of test, transforming them instead into a sort of scientific folk wisdom. When I ask biologists what they think of the idea of key innovations, they

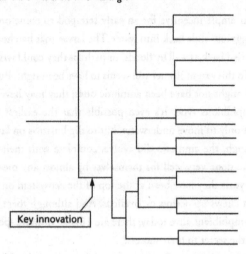

Key innovation

A key innovation encourages a lineage to diversify. In this hypothetical cladogram, for example, the lineage with a key innovation splits into ten species, while the most closely related lineage that lacks it only has four.

have a habit of screwing up one eye and rocking their heads a little from side to side. "I have a hard time with it," says Neil Shubin. "I kind of know them when I see them, ex post facto. But if someone handed me a creature today and said, 'This mutant has a key innovation,' would I really know it? It's not a priori. I wouldn't doubt that they happen, but how do you identify them, and once you i.d. them, what do you do with them?"

Let's say that you think that a key innovation has been responsible for some huge diversity of some group of animals. Perhaps it opened up a new set of ecological niches these animals could never get to before. To put the claim to a test, you can't simply add up how many species are in a given Linnaean family that has the innovation and compare it to others that don't, because these ranks don't reflect the course of evolution very well. A biologist has to look instead at where the innovation arose on a cladogram, add up all the species that lie on branches above it, and compare it to the species belonging to the closest branch that lack it.

And once you've set up a hypothesis this way, you still have to cope with the fact that there are many ways in which a lineage may diversify that have nothing to do with moving into new ecological niches. It may just happen to live in a forest where rivers continually redivide territories and isolate populations.

Even by pure chance, the rate at which its species originate and become extinct may differ from another group's. Biologists would be wrong in such cases to look for some new innovation to explain success. One way to evade these confusing influences is to run several tests instead of just one. Say a certain key innovation has appeared independently in several lineages. If in every case the lineage that possesses it is more diverse than the closest branch descended from a common ancestor, the connection between the innovation and the diversity becomes more persuasive.

In 1996 Robert Reisz and fellow scientists at the University of Toronto tested the old theoretical chestnut that amniotic eggs were the key innovation for amniote success. Their results are tantalizing, but all the caveats about key innovations should be kept in mind. One way of looking at the question is to gauge how the history of amniotes changed after the first eggs appeared. Eggs are so fragile that few manage to become fossils; while the oldest known fossilized egg is about 250 million years old, they are probably much older. It's possible to say this by looking again at a cladogram. One of the oldest known true amniotes is a sleek little 320-million-year-old beast named *Paleothyris,* which left its bones at the base of a petrified tree in Nova Scotia. Paleontologists have recognized that it is a primitive reptile.

Because all of these living lineages lay amniotic eggs, their common ancestor probably did as well. Because *Paleothyris* nestles comfortably in this tree, it most likely laid an amniotic egg as well. And so it's possible to confidently push back the origin of these eggs at least 70 million years before their oldest fossil and probably further back. If the amniotic egg actually did liberate amniotes from

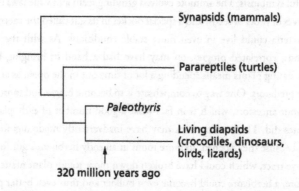

320 million years ago

Synapsids (mammals)

Parareptiles (turtles)

Paleothyris

Living diapsids
(crocodiles, dinosaurs,
birds, lizards)

A fossil called *Paleothyris* pins the origin of amniote eggs to at least 320 million years ago.

water as people assume and let them evolve into all sorts of new forms, then we'd expect that once eggs had evolved, amniotes would start outstripping amphibians in diversity. That's not what happened. For over 20 million years amniotes lived in the shadows, their diversity far lower than that of amphibians. Only 300 million years ago did the diversity of amniote species start a steep climb. Their blooming wasn't part of some general tetrapod golden age: during the amniote explosion amphibians were stagnating at their previous levels—levels they still haven't transcended for over 300 million years.

The Toronto researchers think that the fossils are hinting toward a different cause of amniote success. While amphibian larvae often feed on algae in ponds, you'll never see a frog grazing in a meadow. The fossil record shows that herbivory began independently among several minor branches of small, early tetrapods that had probably evolved to eat insects on land. Rather than crouching in wait and ambushing prey, they had to travel to find their food and needed strong jaws to crush the hard exoskeletons of the bugs they grabbed. As they seized their prey in dense thickets, sometimes they might end up with a seed or other part of a plant in their mouths. Gradually they evolved the enzymes that could digest these foods.

Among the amniotes and their closest relatives, this omnivory shifted to full-blown herbivory. Amniotes snuffling around the ground may have become infected with a number of different strains of soil bacteria, and a few strains that specialized in decomposing plant matter discovered that the intestines of a tetrapod are a pleasant place to be. Once they had broken down the hard plant matter that an amniote might eat, their new host could absorb the sugars and other nutrients through its bowel walls. In time amniote and microbe coevolved into a powerful symbiosis. The amniote evolved grinding teeth and wide jaws to prepare the leaves or roots or seeds, and a special pocket in its gut called the caecum where the bacteria could live in even more stable conditions. As with the origin of tetrapods, correlated progression may have had a hand in bringing herbivory about. Eating plants means spending a lot of time out in the open, as an easy target for predators. One way to compensate is to become bigger and more heavyset than your ancestors, which is in fact what a good number of early plant-eating amniotes did. These adaptations may have inadvertently made digesting plants easier, because they opened up more room in an early herbivore's gut for a bigger digestive tract, which could have broken down more tough plant matter. Able to eat more, a herbivore could become even bulkier and thus even better protected. While an herbivore's stomach doesn't fossilize, its blunt teeth and stocky ribs do, and so it's possible to date the first herbivorous amniotes to around 300 million years ago, which is much closer to when the amniote boom began.

Rather than evolving once, plant-eating actually evolved independently in four different lineages of early amniotes. Other amniote branches continued onward as carnivores, and in the 300 million years that have followed, over a dozen later lineages of meat-eating amniotes have also switched to feeding on plants. The Toronto researchers measured how well these recent converts did compared to their closest carnivorous relatives by looking at the diversity of their living members, comparing, for example, bats that eat fruit and others that eat insects. The researchers studied fourteen groups and found that on average the herbivores have become seventeen times more diverse. Such a strong, repeated pattern suggests that herbivory consistently brings success to those who use it—that it is, in other words, a true key innovation.

In general plants may simply offer more kinds of cuisine than other foods. A tree is made of roots, bark, sap, shoots, leaves, seeds, and—as of about 150 million years ago—flowers and fruits. There are many other kinds of plants to choose from, each with its own menu. Once an amniote evolves the basic equipment for processing plant matter—guts, teeth, and jaws—it can begin to specialize, diverge into different species, and thrive in any habitat where plants grow. And back 300 million years ago, this key innovation's effects probably weren't limited to herbivores alone. Up until that point amniote predators were either small insect-eaters or large creatures still hugging rivers and swamps, feeding on fish and tetrapods still living in the water. But with the evolution of herbivorous tetrapods, dry land became dotted with big, slow, fleshy plant-eaters, which themselves were a new set of ecological niches to which predators could adapt.

Even if the amniotic egg wasn't the special ingredient that made amniotes the dominant vertebrate on land today, it certainly wasn't an obscure little flourish to tetrapod evolution. Some early relatives of amphibians were eating insects at the same time as the early amniotes, and perhaps even eating a little plant matter with it, and yet amphibians never once became full-fledged, lifelong plant-eaters. The difference may have had something to do with their eggs. Amphibians generally drop their eggs in water and leave, while amniotes give them attention until birth and sometimes long afterward. Mingling this way, amniote parents might reliably infect their babies with plant-digesting bacteria. It also appears that you can sometimes be too small to be an herbivore. Among lizards, for example, all of the species that eat plants weigh more than ten ounces. Because amphibians are hatched from their small jellied eggs, they may be unable to become big enough to contain the guts necessary to house that bacteria. Perhaps amniote eggs were the prerequisite for lineages to evolve into herbivores.

Here then is the still-evolving end to the story of how our ancestors came out of the sea. It's fitting that open questions come at its conclusion, because the transition from water to land hasn't really ended. All of us—human, toad, and tern—are still trying to adjust the old lobe-finned body plan to this new life. Living amphibians are not passive relics of the Devonian because we humans didn't evolve from them. They have struggled for 360 million years to survive on land—early amphibians evolved a terrestrial ear millions of years before amniotes did, and today you can find among their descendants countless beautiful new adaptations to land. Some salamanders breathe air not through their lungs but through their skin. Certain frogs coat their skins in an impermeable layer of mucus that lets them trap water, while some salamanders can burrow into the ground during a drought that would kill a rhino and sleep for months. Other frogs can survive a winter frozen solid, while High Yosemite toads have been seen tiptoeing across drifts of snow. The life of a poison dart frog may still revolve around water, but water collected in the interstices of leaves high in rain forest canopies. Not only do they not have to live in a river; their feet never even touch the ground. Amniotes themselves have never been so free of water as we might think. Ponds still draw them back for a drink, they still fish, and some have gone back to sea altogether. And we humans, have we really come all that far from *Acanthostega* in its Greenland stream? The difference is that we sometimes no longer go to the water; we bring it to us. The usual nominees for human achievement include things like words and blades. I say: plumbing.

Chapter Five

THE MIND AT SEA

It's tempting to think that the transition of tetrapods onto dry land was the most spectacular transformation vertebrates ever performed. You need only see a dolphin throw itself from the sea into the air or a pod of finback whales breach and spout clouds of mist from their blowholes to know that it's a temptation to be resisted. We have finally begun to figure out how millions of years of adaptations, of synchronized embryonic rejiggering, of partnerships with bacteria and inventions of things like ears and fingers made it possible for a lobe-fin to live permanently on shore, a possibility that tetrapods are still perfecting. But look at whales and dolphins, how they can swim from Japan to Baja, how they catch more fish than all of our trawlers can gather, how they are as at home at sea as any shark, and how dry land to them means nothing more than death. They are big-brained, perpetual acrobats who spend their lives touring the flooded three-quarters of the world, leaving us humans to our dry, ground-leashed lives. They need none of our plumbing—they get all the fresh water they need from the animals they eat and the vapor in the air they breathe. And yet, despite all this, they are also tetrapods: even mammals like ourselves. Who would dare trace the ancestry of such animals back to land?

For over a century most evolutionary biologists shied away from the challenge because they had so little evidence to work with. In the 1980s and 1990s, however, a loose coalition of paleontologists, geneticists, functional morphologists, comparative psychologists, and embryologists has risen to the dare, and the story these scientists are putting together is so solid that it can serve as a model for macroevolution in general. The rise of whales demonstrates how it not only creates anatomy but erases it, how it cobbles together exaptations into new body parts. It even shows how ultimately macroevolution can create that most elusive feature of life: intelligence. But before plunging down through the murk of time to the origin of cetaceans, it's worth staying for a moment here on the surface, to appreciate exactly what macroevolution has created by considering what a complex work of beauty a cetacean is.

There are seventy-nine known living species of cetaceans swimming in the oceans and rivers of our planet, bound together by some obvious traits that no other animals share. Each cetacean, for example, swims by bending its back up and down, raising a tail with flukes at its tip, and steers with flippers attached to its shoulders. Each breathes through a blowhole and nurses its young at teats hidden in folds of skin. But as similar as cetaceans are in some ways, they fall into two distinct groups. One of these suborders is the mysticetes, or baleen whales. It includes the true cetacean giants, such as the one-hundred-foot, two-hundred-ton blue whale, as well as smaller species like the thirty-three-foot minke. What unites mysticetes as a suborder is their baleen—curtains of horny fronds hanging from the top of their giant, bowed upper jaws. Most species of baleen whales drop their pleated lower jaws and engulf schools of small fish or invertebrates. Closing their mouths, they ram their tongues against the baleen, squeezing out the water through their lips, while leaving the food behind. But some mysticetes trawl in their own fashion: gray whales plow into the ocean bottom and scoop up hundreds of pounds of mud that they sieve for crustaceans, while humpback whales, which hunt schools of small fish like herring, corral them by blasting nets of bubbles out of their mouths and blowholes and then wheeling around to sweep the fish up.

The other suborder is called the odontocetes, or the toothed whales. It includes dolphins, porpoises, sperm whales, and killer whales, as well as more exotic species like the beaked whales and narwhals. Most toothed whales have long rows of uniform pegs for teeth, but some break the rule; the narwhal has a single unicorn's stake in its mouth, while most beaked whales have a tooth shaped like an ax blade on each side of the lower jaw. All odontocetes use a biological sonar: by emitting high-frequency calls and listening to their echoes, they can perceive the ocean with three-dimensional clarity. They use their sonar to hunt—sperm whales use it to chase after giant squid, dolphins after small fish like herrings.

How animals get their food helps to shape their societies, and cetaceans are no exception. Many baleen whale species mingle casually on feeding grounds and then migrate individually to another region where they meet again to mate. Toothed whales, on the other hand, often form big societies that last for decades and hunt, like wolves or humans, in cooperative packs. Most of the details of cetacean social life are still a mystery, though, because it's so hard to watch whales in the wild. The beaked whales keep so faithfully to the open ocean and deep waters that we learn of a new species only when it happens to beach on some remote shore. No one has ever seen a living giant squid, let alone one in combat with a sperm whale.

Even if cetologists were to spend enough time at sea to decode the complete social life of whales, they still wouldn't be able to answer other questions, such as how they swim, how they survive underwater, or how they think. That kind of knowledge can only come from experiments on captive cetaceans. Yet no one would bother to try getting a full-length sperm whale into an aquarium tank. Even many of the smaller porpoises and dolphins do badly out of the ocean and away from their normal social life. Of all the cetaceans, in fact, only one has let humans get acquainted with it in all possible ways—physiologically, hydrodynamically, anatomically, and psychologically: the bottlenose dolphin. The bottlenose is also the cetacean the rest of us know best—it's Flipper, it's the animal that packs people onto bleachers at aquariums for performances, that stars in movies, that models for glass figurines and earrings, that turns up in New Age psychotherapy and midwifery. Only a few animals have become standards for scientific study, such as fruit flies, rats, zebra fish, and nematode worms, but bottlenose dolphins are the only ones who are adored by anyone other than a postdoc.

When a bottlenose dolphin is swimming, it looks as if it is making no more effort than a thought, whether it sprints along the surface or dives six hundred feet underwater. When you watch one, it's hard to keep in mind that it is a mammal, with a fast-burning, warm-blooded metabolism that has to be stoked with oxygen. A bottlenose doesn't call attention to the fact that it swims underwater while holding its breath; every few minutes it will casually break off whatever it happens to be doing and make a few quick upward kicks to reach the ocean's surface.

In order to be a graceful swimmer an animal first needs the proper shape. If an object moving through water has a hydrodynamically ugly form—a block of wood, for example—the water smashes into its broad face and is pushed to its sharp edges where it breaks away and spins off into eddies. These eddies draw off the block's forward energy to feed their own swirlings, and the block soon glides to a stop. But if an object is long, rounded, and tapered—a dolphin, in other words—the water will flow along its contours smoothly. The shape of a dolphin won't allow it to glide frictionlessly forever, but it will need far less energy to swim than a cubic cetacean.

A dolphin finds that energy for swimming in its tail. Extending off from either side of the end of its spine are two long flukes made of connective tissue that taper to long, winglike tips. Their hydrodynamics actually work almost exactly like a bird wing's aerodynamics. When a bird pushes its wings down, keeping its profile slightly tilted upward, the air flows over it in an unbalanced way that creates a force directed up and forward. By working the muscles along its back and sides, a dolphin pulls down its tail, its fluke in the same profile, and creates the same effect. On a bird's upstroke, however, the same movement would push it forward and down, and in its endless fight against gravity it can't afford to drive itself toward the ground. Instead it tucks in its wings as it raises them. A dolphin has no such worries about crashing into the sea floor: because it is almost neutrally buoyant it travels as if it were an astronaut in space. It can raise its tail and propel itself farther forward, something like a reversible bird.

The serious study of how dolphins swim is a little over sixty years old. In 1936 Sir James Gray, then the world's expert on animal locomotion, studied a film that showed a dolphin shooting past a 136-foot-long ship in the Indian Ocean in seven seconds—a speed, he calculated, of twenty-three miles an hour. Gray made some standard calculations of the drag that a dolphin would incur swimming at that speed and how much power it would have to supply in order to overcome it to keep moving forward. It turned out to be seven times more than normal mammal muscle can produce, in an animal that for the most part held its breath. The only way Gray could imagine resolving the paradox was to accept that dolphins kept their drag low by keeping the flow of water far smoother along its body than physics would predict.

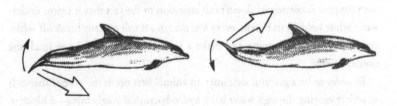

In both its upstroke and downstroke, a dolphin propels itself forward.

Gray's Paradox, as it came to be called, baffled later biologists, and when the cold war began, the American and Soviet navies turned their minds to it as well. They thought that if they could solve it they could design submarines that would slip through the water efficiently and silently. Perhaps the heat coming from a warm-blooded animal made the surrounding water less sticky and less prone to break into eddies. Maybe dolphin skin had ridges that could channel the water down their flanks. Maybe its gooey eye secretions coated its body. The American navy invented rubberized paints on the suspicion that the rubbery skin of dolphins could damp out little waves that could become the seeds of turbulence. Soviet scientists even went so far as to drag naked women through flumes to see how their fat rippled in water compared to dolphins. But no one could find the secret of dolphin swimming.

These days, although a few researchers in Russia still struggle on, most hydrodynamics experts think that solving Gray's Paradox is a fool's errand. Gray had relied on measures of muscle power that had been taken from men rowing for fifteen-minute bouts. In seven seconds it's possible for a mammal to release a far greater burst of power. While dolphins can swim twenty-three miles an hour, they prefer speeds between three and ten miles. Dolphins have no great hydrodynamic secret beyond their obvious shape, and they still have to cope with turbulence and fairly high drag forces. Without any miracles from their skin or their geometry, they have to generate tremendous forces to muscle through the water. Rather than a secret that could win the cold war, dolphins possess an ordinariness that makes them a puzzle.

Of course even the most exquisite hydrodynamic shape is worthless if an animal has no way to propel it. When physiologists look at swimming bottlenose dolphins, they often wonder if the animals have some exotic way to burn energy efficiently. Only recently, though, have they begun to measure dolphin physiology reliably. They use monitors attached to dolphins by suction cups to capture their heart rate, and from that figure they can calculate how much oxygen they are burning. By comparing the oxygen to the speed at which the dolphins swim, the physiologists come up with a figure that's known as a "cost of transport." Technically, this is milliliters of oxygen consumed per kilogram of body weight per kilometer. Untechnically, it's akin to a car's miles per gallon, reflecting how much energy it takes to move a given weight of animal a given distance. Physiologists have found that a dolphin has the cheapest cost of transport of all swimming mammals, only twice as much as the cost of a fish of equivalent size. Humans have one that's over twenty times higher. Yet a swimming dolphin's efficiency becomes much less exceptional if you compare it to terrestrial mammals running on land—judging how well each animal moves in its own element. In such a lineup, a dolphin is actually about as impressive as an ox.

Cetologists haven't yet completely figured out how a dolphin can swim so well with the hydrodynamics and physiology that it has. Part of the answer is that dolphins use many little tricks to move efficiently through the water. Wild dolphins are famous for surfing on bow waves of ships, sometimes just below the surface or bounding along on the crests. It may look like they are doing it for pleasure, but surfing makes sober, cost-cutting good sense: a dolphin surfing at eight miles an hour has the same cost of transport as it does swimming at five miles an hour. Most likely dolphins have been grabbing cheap rides since long before the first boat cut a wake, on waves churned by the winds or big whales. The surface isn't the only place dolphins cheat: on deep dives their rib cages become compressed at about 150 feet down and their lungs squeeze shut. They become negatively buoyant and sink like stones. It seems that the dolphins themselves are aware of this trick, because when they lose their buoyancy they stop swimming until they reach the depth they want.

Other tricks are anatomical. As mammals, bottlenose dolphins need to keep their bodies at a constant, high temperature—a hard feat to manage in water, which draws heat quickly from an animal in all but the most tropical latitudes. Dolphins, like all cetaceans, keep the heat in with a thick layer of blubber. They also almost never stop moving from birth to death—they even swim while they sleep, with only half of the brain shut down at a time—and the endless work warms their bodies. In fact, dolphins have to worry as much about overheating as losing heat.

We terrestrial mammals also risk overheating from time to time, and in order to dump this extra heat we use a special set of blood vessels. They branch off of major arteries in our arms and legs and travel just underneath our skin, where the heat they carry rushes quickly into the air. If we become too chilly, though, our bodies shut these vessels off at choke points, making our fingers and toes cold but conserving the warmth in our core. A dolphin has similar vessels; they poke through its blubber layer and run underneath the skin of its flukes and fins. After a hard swim, it opens the vessels and dumps the heat in a blaze.

Overheating poses threats to dolphins that we never have to face on land. Swimming, for example, should make a male dolphin sterile. Sperm can only develop and survive at temperatures a few degrees below that of a mammal's core, and so most males keep their testicles in a sac that hangs down from the body. Clever as this arrangement may be on land, a swimming animal does not want a bag hanging from its streamlined body, and so male dolphins lodge their blimp-shaped testicles snugly in their body. Now they can swim quickly, but they can't have children because the testicles sit between massive muscles that work continuously as they swim, and are fed by vessels coming off the nearby aorta, full of hot blood. The arrangement makes as much sense as trying to keep a tub of ice cream cold by putting it on an engine block.

A male dolphin preserves his fertility by rerouting his circulation. After blood travels to his tail and fin and unloads its heat, it flows through veins directly to his gonads. There the veins split up into fine capillaries that run side by side along the arterial vessels, cooling them, and the arterial vessels in turn chill the dolphin's sperm. A male dolphin's testicles are 1.4 degrees Fahrenheit cooler than the surrounding body, and the refrigeration is so powerful that when the dolphin takes long, fast swims—heating its body even more—the testicle temperature actually drops another half a degree.

Female dolphins need just as badly to keep their gonads cool. A mammal fetus burns like a little furnace inside its mother, its metabolic rate running two or three times higher than that of the mother's tissues. That heat has to get out of the womb somehow or the baby will be born deformed or dead. A pregnant woman carries off some of the warmth through her blood vessels, while the rest gets conducted away through her belly. But a dolphin's uterus is in the same neighborhood of its body as the testicles, which means that between the fetus and the ocean is a hard-working layer of abdominal muscle producing heat of its own, and beyond that a layer of insulating blubber. Without an abdominal window, the fetus should burn up. But like males, female dolphins have a circulation system that lets them pump blood from their tails directly to their wombs to keep their babies chilled.

Underneath their tapered skin and their maze of veins and arteries is the anatomy that propels bottlenose dolphins in the first place—their skeleton and muscles. I once watched a dolphin autopsy, and the most startling moment for me was when the biologists had stripped the spine bare. The biggest bones and muscles in humans (as in any land mammal) are in and around our limbs, because they carry us forward. Our backs need to do little more than keep our guts and heads aloft. Dolphins move only with their backs, though, and looking at the autopsied skeleton I could see its pure dedication. Each vertebra was crowned with a huge upside-down T of bone, and on the two shelves it formed rested long snakes of muscle that ran from the dolphin's head to its tail. As I looked at its skeleton I reached my hand around my own back, and I touched the feeble chain of bumps on my spine.

To swim, dolphins do not simply contract these long muscles and bend their backs. From the back of a dolphin's head to the base of its tail, a sheath of barely visible connective tissue wraps around the muscles running on top and on the sides of the spine. It is cross-woven like a sleeve of fabric, but it doesn't merely encase the muscles like the casing of a sausage. Muscles along the length of the dolphin are anchored to the sheath, and it is lashed at points to the spine itself. Ann Pabst, a biologist at the University of North Carolina, needed five years to map out the shape of the sheath, and her work suggests that when a dolphin wants to bend its back, one small set of muscles first contracts, stiffening parts of the sheath until it becomes as rigid as a bone. A second set of muscles anchored to these rigid parts of the sheath can then use it as a sort of substitute spine to help them transmit their forces down the length of the dolphin's body. Just short of the flukes, the muscles attach again to the sheath, which projects tendons down to the tip of its tail, and these tendons can tug the flukes up and down to the best angle for generating lift.

It's possible that the sheath may also act like a spring. Basically, a dolphin is a pressurized cylinder wrapped by a sheath of helically wound fibers. This kind of shape has some valuable properties that animals as varied as squid, sharks, caecilians, and earthworms take advantage of. Such a cylinder won't kink when it bends, and the angle of the fibers lets it resist twisting. If the fibers are oriented at a certain range of angles, the cylinder becomes so springy that it bounces back if it's bent. In other words, the energy that goes into the bending is stored in the fibers and then released again.

If dolphins could save energy this way, it would explain some strange data that have emerged in the past few years. As dolphins swim harder and harder, their oxygen consumption smoothly increases and then levels off. This plateau isn't surprising, since animals have limits to their aerobic metabolism; to push any harder they have to resort to other metabolic chains of chemical reactions that don't use oxygen. For humans, as with most other animals, this anaerobic energy is short-lived because it generates lactic acid as waste that builds up in the muscles, creating a heavy, achey feel. But dolphins can generate higher and higher forces while building up only paltry levels of lactic acid. Only one other mammal can compare: the kangaroo.

Kangaroos can continue hopping far beyond their aerobic capacity because they have massive springs in their legs. With each jump, they stretch the stiff strands of collagen in their tendons and store much of the energy of their fall; when they spring out for the next hop, the tendons return 92 percent of the energy put into them. When a kangaroo wants to hop faster than the speed its aerobic metabolism alone can provide, it uses the springs in its legs rather than lactic acid–generating chemistry. Dolphins may be doing the same thing in water. Perhaps at a certain frequency the upstroke of a dolphin's tail can load energy into the subdermal sheath, which springs back, helping to push the tail down through its downstroke. At the right frequency, dolphins would resonate as they swam, ringing like bells. If they did, it would make a little more sense how such an amazing swimmer manages with the physiology of an ox.

Researchers have speculated in the past that dolphins might swim with springs, but their results were ambiguous. The trouble is that cetaceans are not simple cylinders; their tails narrow and flatten before spreading into flukes, and the sheath does confusing things like anchor to the spine. Still, the angles of the fibers in the sheath are correct for making the tail vibrate like a spring. The best way to test this idea would be to measure the resilience of the sheath, but it turns out that the stretching machines that engineers use change the orientations of the fibers as they pull, making it impossible to get an accurate reading.

Ann Pabst has had more luck with blubber. Blubber is not, as you might imagine, a blanket of uninteresting fat. You can see the difference with a microscope using polarized light: the fat of a cow appears as an expanse of purple blobs, but blubber contains gorgeous blue and gold stripes of connective tissue woven into the blobs like a Japanese tapestry. Here is a second cross-woven material sitting on top of the sheath and encasing the dolphin, with fiber angles that match those of the sheath underneath. But unlike the sheath, blubber holds its shape when it's clamped and stretched. Resilience is just a measure of the amount of energy you get out of a system relative to the amount you put in. While collagen—the finest biological spring—has a value of 92 percent, blubber has a respectable value of 87 percent.

Swimming demands a complicated anatomy—complete with refrigerators, springs, sheaths, and levers—but inside the forehead of a dolphin are organs so strange and intricate that they make the rest of its body seem as easy to comprehend as a wheelbarrow. Hidden in the dolphin's head is the ability to see the ocean with sound. When Darwin wondered out loud about how evolution could create complex structures, the example he chose was the eye. He probably could not have thought of a more intricate, sensitive organ than this little globe that lets us see only because so many different parts—lens, pupil, retina, vitreous jelly, muscle, pigment, blood, nerve—work together to focus light into images. It might seem impossible for them to have all simply fallen blindly together into such a sophisticated combination. If he were writing now, I think Darwin might choose the sonar equipment of a dolphin.

The best guide through the cogs and gears of this device is Ted Cranford. Cranford has been studying the insides of whale heads since 1984, under the guidance of Ken Norris, a now-retired professor at the University of California at Santa Cruz and guru of the echolocation set. In 1960 zoologists knew that toothed whales made clicks, buzzes, whistles, and pops and that they could use the clicks to detect objects around them by their echoes, as do bats. It was naturally assumed that the sounds came from the odontocetes' throats. When an anatomist cuts the tongue out of a dead dolphin's mouth, it looks like a fleshy Valentine's Day heart, and sprouting from the back of the tongue is a long stalk that ends in what looks like a pair of smiling lips. This is the larynx of the animal. Given that the larynx in mammals like ourselves creates sound, and that in the dolphin it is controlled by eighteen individual muscles, most people thought that it must be where a toothed whale's voice came from as well. But in 1960 Norris defied this common sense by pointing out that a dolphin can't detect an object underneath its jaw. In fact it was able to locate an object most accurately if it was near the upper quadrant of the dolphin's smoothly curved head. Norris claimed that sounds therefore could not be coming out of a dolphin's throat—they must be emitted from its nasal passage somewhere between the point where it emerges from the skull and where it ends at the blowhole.

In the years that followed, experiments have confirmed that Norris was right, but none so much as Cranford's work. When Cranford first began to study at Santa Cruz, Norris suggested that he reconstruct the head of a dolphin inside a computer. Anatomists had been doing dissections on dolphins for decades, but as important as their work was, the methods had limits. "All you got in anatomy papers was the typical knee-bone-is-connected-to-the-thigh-bone kind of writing," says Cranford. The equipment of echolocation—sacs, tubes, bits of fat, bones, gristle—is arranged so intricately in all three dimensions that it can't be understood by merely cutting up a dolphin's head. Someone needed to see the structures in their natural position in order to say if Norris really was right. Without this kind of detail, no one could know exactly what was making the sound and where it was going. Cranford thought the project would be straightforward enough—he would simply do what Erik Jarvik had done with his lobefinned *Eusthenopteron:* make slices of the head and photograph the outlines of the structures inside each layer. He would then scan the pictures into a computer and use it to calculate the three-dimensional shapes inside its head.

Cranford got hold of the head of a bottlenose dolphin that had stranded and died on the California coast. He froze it solid, with a plan to slice it into hundreds of pieces with a diamond-wire saw and photograph each one. It took him a year to scrounge up the money to rent a saw and buy special cutting wires beaded with diamonds. "The only way I could get the dolphin head really hard was to set the head up in the freezer. So I set the saw up in a freezer at Sea World one summer. Here I am working with skuas and penguins running around my feet, and it's like I'm in the Arctic. Things went badly. The tissue was frozen solid, but it was still too soft and it was stripping the diamonds off the wires. This guy was making me special diamond wires with three times as many diamonds on them—they were $350 each—and here I was stripping diamonds off them."

After a month Cranford abandoned the freezer, pretty sure his project was an expensive failure. But before he had begun sawing he had made a series of CT scans on the off chance that they might be able to capture some detail. He looked them over and realized that he had drastically underestimated the power of the machine: he could see in the scans the bone, sacs, fat, gristle, and muscle of the dolphin's head, all gorgeously detailed. He could project the slices onto a screen, draw the outlines of the blobs and curves, and turn them into a computerized head. The results were so good that he was able to use CT scans to study other odontocetes, including pygmy sperm whales, harbor porpoises, franciscanas, and Pacific white-side dolphins. The days of cold despair and diamonds lost are now far behind: Cranford can at last guide people along the path that he believes sound takes out of a dolphin's head and back in.

It starts with the larynx. Acting like a pump, it rams into the nasal passage that runs from the dolphin's throat to the top of its skull. Because the bony walls can't bend, the pressure of the air squeezed in the cavity soars. The nasal passage emerges from the skull as two fleshy tubes side by side. A little of the air flows into a set of sacs that branch from each tube, but most of it presses fiercely at the top end of the tubes. They are sealed by flaps of fat and gristle, whose pinched, bare texture inspired nineteenth-century anatomists to name them monkey lips. Carefully controlling the monkey lips with its facial muscles, a dolphin can let the trapped air escape between them, making the monkey lips vibrate like the lips of a trumpet player. After air has escaped past them, the dolphin doesn't let it slip out its blowhole and go to waste; the blowhole stays tightly shut and the air flows instead into two sacs surrounding the nasal passage. The dolphin squeezes the air back out of the sacs and back down through its nose, so that it can be pushed back up again between the monkey lips.

It is not the air itself, however, that makes the noise. Airborne vibration is useless underwater: water is so much denser than air that airborne sound will bounce away from it like light hitting a mirror. But vibrations made in watery, soft tissue—particularly certain fats—can move almost effortlessly out into water because the substances are so acoustically alike. The air passing between a dolphin's monkey lips makes the lips themselves vibrate up to thousands of times a second, and this, Cranford thinks, is the source of a dolphin's underwater sounds.

Lodged in the lips are dollops of fat called the dorsal bursae, and the bursae are connected to a large piece of fatty tissue that gives a dolphin its bulging forehead. Called a melon, it nestles on top of the dolphin's concave skull. Once biologists thought that the melon was a sort of bumper that only helped make a dolphin's body more hydrodynamic, but it's now considered an acoustic lens that can change the path of the signals. "These fats are poisonous to the metabolic machinery," says Cranford. "In other words, they're not there for energy storage, and they're not there for trivial functions like being a hydrodynamic bumper. Those are pretty expensive molecules to use for that purpose—you could use connective tissue instead." As they make sounds, toothed whales can control the shape of the melon with their facial muscles, possibly shifting the paths of the sound waves.

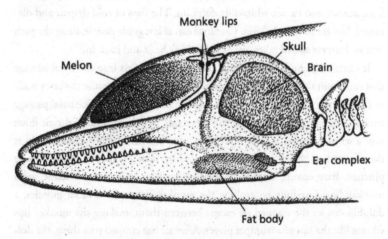

A cross section of a dolphin's head. Air passing through the nasal passage makes the monkey lips vibrate, and these vibrations pass through the melon and out into water.

Tracking the sound from the monkey lips through the melon and out of the dolphin's head is still beyond the abilities of scientists such as Cranford. Some sound waves move backward and bounce off the skull, while some reflect off the many air sacs around the nose—and bone and air are completely different sorts of mirrors. Each nasal tube has its own set of monkey lips, and a dolphin may be able to work them independently. The melon's fat is made of molecular chains of many different lengths, and each changes the properties of a sound in its own way. These fats seem to be distributed with a distinct kind of topography throughout the melon, but no one knows its effects on the pulses. All these sound waves also interfere with each other, reinforcing in some places, damping out in others, changing not only the tone of the sound but its direction.

"People don't think about this if they're trained as engineers or physicists," says Cranford. "Their training requires them to distill a system down to its simple elements, and they can go a long way with that. But a biologist is trained to see things as complex. If you try to solve a partial differential equation, and put in all the conditions you find in a dolphin's head, then you have places where the boundaries are hard, places where they are soft, places where they are soft but highly reflective—it goes on and on. Even the most powerful computers will have a tough time chewing that up."

The best that cetologists can do for now is study the sounds as they come out of a dolphin head. The buzzes, whistles, and pops seem to serve as communication between dolphins, while the clicks reveal their surroundings. They emerge from the melon as a beam of sound, a beam that dolphins can somehow scan like a flashlight through the water. And like photons in a ray of light, the sound waves reflect off objects they hit. In order to hear so sensitively, a dolphin needs a very different anatomy from a mammal on land. You can tell that a mockingbird is singing to your left because its airborne song cannot penetrate bone. The sound travels directly to your left eardrum through the ear's opening but has to bounce off other objects to make its way around your head to reach the right one. There's a slight delay between the two signals, and your brain uses it to figure out where the mockingbird is. But when you swim underwater in a lake and hear a motorboat passing overhead, you won't be able to tell where it is. If the motorboat is on your left, its sound travels through the water and hits your left ear directly. In order to reach your right ear it doesn't have to travel around the circumference of your head as it does in air. Because some underwater sound can penetrate bone, the vibrations from the boat simply travel the much shorter path through your skull to your right ear, confusing your delay-measuring brain.

A cetacean is spared this confusion because its ear is separated from its skull. Sitting in a bony case shaped like a grape, it hangs down just behind the jaw, barely connected to the skull by a single flange of bone and a few ligaments and surrounded by pockets of air and foam. Insulated this way, the ear cannot pick up stray sounds passing through the whale's head, including its own intense echolocating blasts. Yet a dolphin doesn't actually hear through the openings to its ear. Its jaw is hollowed out and loaded with the same kind of fat that makes up its melon, which means that incoming sound travels more easily through the fat than through the bone, working its way back to the ear abutting the jaw. It acts like an acoustic pipeline, pumping sound to the dolphin's ear, which contains a forest of hair cells six times denser than in the average mammal's, which transfer the signals to the brain.

As awkward as it might seem for a toothed whale to blow its nose and listen for the echoes with its jaw, odontocetes can perceive the underwater world with a clarity we can only dream of. They are like living ultrasound machines, seeing the interior of the animals around them. Researchers have been able to gauge the perceptions of dolphins by training them to touch a paddle when they hear an object put in front of them. A dolphin can shriek at a tangerine-sized steel ball over a football field away, receive the echo (which is a million times weaker by the time it returns), and tell a human that the ball is there. It can tell the difference between a sphere and a disk. It can distinguish between cylinders that differ by a fraction of an inch, and if researchers try to trick it by using two cylinders of the same size instead, it will simply refuse to respond. It can hear the difference between copper and aluminum.

The perceptions that a dolphin can manage with echolocation demand more than sensitive ears. Their brains must be able to compare the timing of what they hear, and so they can distinguish between sounds separated by twenty-five millionths of a second, forty times briefer than a human brain can. If a dolphin is blindfolded and presented with a pyramid at different orientations—or a cube, or a rectangular block—it can identify the different views as belonging to the same object. Researchers have had dolphins look at intricate constructions of plastic pipes with their eyes and then successfully pick them out from a pipe lineup while blindfolded. Both experiments show that a dolphin not only perceives objects as sharply as we see, but can build abstract, multidimensional representations of them that are not tied to a particular sense.

Nor do their representations even have to be of a physical object. At the University of Hawaii Louis Herman and his coworkers have invented a sign language that they've successfully taught to dolphins. They teach it to the dolphins by example, rewarding them when they respond to a signal correctly. What's most remarkable about Herman's work is that the language is not merely a set of labels. If he gestures to a dolphin the signs RIGHT HOOP LEFT FRISBEE FETCH, it will go to a Frisbee sitting to its left, put it in its mouth, and carry it to the hoop to its right. The Hawaii research suggests that dolphins think grammatically and can handle abstract representations not just of things but of their relationships to other things. Once the researchers instructed a dolphin to swim through a hoop, but the hoop was lying flat on the bottom of the pool. The dolphin swam down, snatched the hoop, and put it in a position where it could carry out the instructions.

Like human children, the dolphins learn this language by example rather than rule, and yet they come to understand the implicit grammar so well that they know when it is being violated. In Herman's transpecies language, for example, PERSON WATER HOOP FETCH makes no sense. When human subjects are taught an artificial language and then get a nonsensical command, they respond by extracting what little sense they can. That is what dolphins do as well: in their minds they delete the word WATER from the command, reducing it to PERSON HOOP FETCH, and bring the person the hoop. If the sentence makes no sense at all—PERSON WATER FETCH—the dolphin will wait patiently for the human to stop babbling. And most remarkable of all, a researcher can tell the dolphins simply to do something new—perhaps some routine with one of their toys or perhaps some acrobatics they've never tried before. Two dolphins can be told to do something new together and they will immediately rush out into the middle of the pool and perform. How they decide to divide the creative labor will have to remain a mystery until we can understand the communications of dolphins.

In the 1960s, when the first work on dolphin intelligence came out, all kinds of wild claims immediately followed—that whales are more intelligent than humans, that in a matter of a few years we would be having philosophical dialogues with a bottlenose. Yet in thirty years work such as the kind done at the University of Hawaii is the best that we've managed. The problem is not that dolphins are dumber than we thought, but that our anthropomorphism inevitably makes it hard to understand an intelligence other than our own.

Scientists like Cranford depend on professional dolphin trainers in order to do their experiments, and the trainers once made him acutely aware of the gulf they face every day. "At parties people sometimes play the Training Game. It's a neat thing. The dolphin-type person goes out of the room, and then everybody else decides what they want the dolphin to do. Stand on their hands and clap their feet three times, or take a ball and put it on someone's head. The dolphin person comes back in the room and the only way you can communicate with this animal is to say 'Go,' or you can blow the whistle and say, 'That's it,' and they come to get a reward.

"So you say 'Go,' and the trainee starts wandering around the room. When they get close to something, you say 'That's it!' So they think, 'Oh, there's some-thing over here.' After they get their reward, they go back over there, and by a series of rewards you train them to do what you want them to do. When I was the dolphin it became blazingly clear that I couldn't figure out what in the heck they wanted me to do. I was doing all kinds of things. You might think they want you to spin around before you pick up the ball, but that wasn't the case—it came about because you happened to spin around and pick up the object be-fore they blew the whistle. And now the trainer thinks I *have* to spin around."

Skilled trainers know how to shave away much of this superstitious behavior, and the hundreds of experiments that they've helped cetologists run are testi-mony to the kind of conversations they can set up with these creatures. But in another sense, the Training Game also captures how hard it is to get inside a dolphin's mind. The performance Cranford settles on at his party barely resem-bles the way he'd do it if he knew exactly what his trainers wanted. Perhaps af-ter a few days of work, the trainer might get him to do the routine properly. But anyone who would try then to explain Cranford's psyche based on his awkward show would do him a grave injustice.

One example of this ambiguity in dolphin psychology is the ongoing re-search into the question of whether dolphins have self-awareness. In the classic test on animals, psychologists see if they recognize themselves in a mirror. Since they can't ask the animals if they recognize themselves, the researchers instead use a roundabout procedure: They first place the mirror in front of the animal for a few days and then remove it. Anesthetizing the animal, they put a mark on its head and put the mirror back. If after seeing itself in the mirror the animal reaches to the mark on its own face, it most likely considers the reflection not another animal but itself. Chimps and orangutans pass the mirror test, while only a single gorilla has managed so far.

A researcher at Emory University in Georgia named Lori Marino undertook the first mirror experiment on dolphins in 1993. She and her colleagues tested two male bottlenose dolphins named Pan and Delphi in a pool with a covered mirror at one end. Each day the dolphins came to a station at one end of the pool and trainers stroked their flanks or painted a spot of blue medicinal liquid that is almost invisible on the dolphin's body. Every two or three days the mirror was unveiled and Marino filmed their behavior. On the eleventh day, during their daily stroke she marked them with a dab of white zinc oxide.

In the first ten days, Pan and Delphi seemed to behave differently in front of the mirror. They would tilt their bodies in front of it over and over again, or circle their heads or open their mouths. Marino also noted that their sex life changed. Pan and Delphi, like most male dolphins, were a randy pair, and every few minutes one would try to penetrate the other. When Marino opened the mirror, they would have sex exclusively in front of it, and if they swam out of the mirror's reflection, the sex promptly stopped. And finally, on the eleventh day, after she had marked Pan with zinc oxide, he and Delphi swam madly around the pool for an hour. She called them back to wipe off the mark, and as soon as she had finished, Pan darted to the mirror before being given his release command, positioned himself where he could see where the patch had been, and then came back to Marino. The same sequence happened with both dolphins three times in a row.

And so are dolphins self-aware? The results won't let Marino come to a definitive conclusion. Everything that Pan and Delphi did that could be construed as a sign of self-awareness could have another explanation. When they opened their mouths, they might have wanted to look down their own throats. But an open mouth is also a display of aggression, and they might have been using it against what they thought was another dolphin. Since dolphins love to imitate one another, there was no way of knowing if these movements were of an animal who was aware of himself, or of a mimic. As for the sex, Pan and Delphi might enjoy watching themselves in the mirror—or they may like having sex in view of other dolphins. As for the mark, the dolphins only positioned themselves in front of the mirror after the mark was gone, not when it was applied.

A choice between self-awareness and the lack of it may be one that dolphins don't have to make. Harry Jerison, a neuroanatomist at the University of California at Los Angeles, has tried to tackle the question of dolphin intelligence by imagining how dolphins perceive their world. What we consider reality is very much a matter of what our senses supply to our brains. Bees can use polarized sunlight as a compass to guide them to flowers and back to their hive, and so a sense of north, south, east, and west probably feels as all-surrounding to them as our perception of the three dimensions of space itself. We perceive the world in relation to our own bodies, our own selves, but to a bee there may be two poles: the bee's self, such as it is, and the sun.

A dolphin gets most of its perceptions from its eyes—a hazy, 360-degree view—and its echolocation, which superimposes a flickering three-dimensional image on it. But dolphins, Jerison points out, may perceive the universe much as humans do. Thanks to language, we can organize our thoughts into a sophisticated representation of the world and the people in it, and we can relay our perception of the world to others by speaking or writing. Hearing becomes not just a way to detect the sound of a falling tree; it becomes a telepathic sense.

Dolphins may also be telepathic in Jerison's sense of the word. As dolphins travel together in tight bands, the shriek of one member echoes back with its information to them all. No individual has any perception that another dolphin doesn't share. It's also possible that dolphins may communicate by manufacturing echolocated visions. If dolphins are in fact continually sharing and exchanging interior and exterior worlds with one another, our notion of self would be meaningless to them. Certainly dolphin societies have hierarchies, conflicts over mates, and other marks of the societies in which individuals struggle. Dolphins may even be able to name each other with signature whistles. But their society may nevertheless be one of an overlapping network of minds, wandering linked through a transparent ocean.

This, from fluke to brain, is a short summary of what we know about the bottlenose dolphin. Perhaps if we can restrain ourselves from slaughtering other cetaceans, we'll be able to learn as much about them. They will have a lot in common, but there will be differences as well. Baleen whales may or may not be able to use a blubber spring as dolphins can, but their huge size may allow them to ride enormous underwater waves instead, supplying them with a third of the energy they need to swim. Baleen whales don't seem to be able to echolocate, but no one has any idea about their perceptions or intelligence. Yet even with what little is known, it's clear that the origin of cetaceans was one of macroevolution's greatest productions. When you consider the fact that it adapted a warm-blooded, air-breathing mammal so well to the ocean, gave it gifts like baleen and echolocation, and endowed it with such a sophisticated, mysterious mind, it becomes obvious that in the history of life there are many stories worth knowing other than our own evolution. And as we start to descend down from the surface of time, toward the transformation that created cetaceans in the first place, we know what a long trip it will be.

Chapter Six

THE EQUATION OF A WHALE

One of the things that impressed Balzac most about Cuvier was how he "rebuilt, like Cadmus, cities from a tooth." The Baron of Fossils earned his reputation with teeth in 1804 when he received the skeleton of a creature the size of a cat from the limestone quarries of Montmartre. It had left its mark on two slabs that fit together like a muffin and its tin. The hips and legs and bits of spine were lodged in one piece of the limestone, while the rest of the skeleton lay in the other: a shoulder, an arm, and part of its jaw, its teeth half-grinning out of the rock. With no clear idea of what kind of animal this was, Cuvier brought the fossil into his museum laboratory and began looking down into the animal's mouth, chipping the bones free and sketching the details as he went.

He could see the bump on the rear of the jaw where it hinged with the skull, a feature called the condyle that only mammals have. In many mammals the condyle rides high on the back of the jaw, but the creature in Cuvier's limestone had a low one that bulged from the jaw barely above its rows of teeth. He could eliminate mammals like cats, dogs, and martens all at once, because only animals such as moles, hedgehogs, bats, and opossums have such a jaw. Cuvier prised the lower jaw free and exposed both sets of teeth. They were neither sharp-edged like a carnivore's nor flattened like those of a grass-chewing cow. Instead they bristled with small pointed cusps—again, the sort shared only by mammals who have a low condyle. Under a magnifying glass, he could see that some of the teeth were triangular, with three hooklike cusps. A mole has seven of these cusps; so does a bat; a hedgehog has four. The only mammals that have three are certain marsupials: the opossums of North and South America and their relatives, the dasyurids of Australia (a group that includes the Tasmanian devil).

"I stopped my work on the teeth before I occupied myself with the rest of the skeleton," Cuvier wrote later, "but I could have predicted everything else from this sole index. Number of parts, forms, proportions—all this the surface of the rock offered us, was found entirely answered in that first sight."

A French opossum would have been a scientific outrage. Opossums are marsupials, and as such they give birth to young the size of rice grains that creep into their mother's pouch and clamp onto a nipple for months. But all living mammals in Europe are placental—they gestate in the womb, nourished by a placenta, and emerge much larger than marsupial babies. There are no living marsupials in Europe, and opossums and dasyurids live only in Australia and the New World. Cuvier had just began to introduce Europe to the concept of extinction, and people could still be skeptical because Cuvier had found only a few fossils that were out of place. If he was right about these new slabs, the skeptics would have a much harder time doubting him.

Cuvier had no doubt in his prediction, though, because he was convinced that the animal's anatomy was not a patchwork. Each organism is born dedicated to its own manner of life, he argued, and all of the parts of its body point toward that end. A mole's backhoe hands and its thick-walled head are not some coincidental arrangement—they are both necessary for digging tunnels underground. And because animals are so unified, their teeth can serve as miniature portraits. "Every organized creature forms a whole, a unique and closed system, whose parts mutually correspond to one another," Cuvier wrote. "In a word, the form of the tooth leads to the form of the condyle, that of the scapula to that of the nails, just as an equation of a curve implies all of its properties."

Of course Cuvier didn't know the equation of an opossum, but he could come close by studying how the parts of a body correlated with one another. With enough observation, naturalists might approach the rational laws of physics and mathematics. Cuvier badly wanted to put anatomy on the level with those sciences, to make it more than a genteel hobby of collecting—so badly that he used his alleged opossum to put on a performance. Having predicted its identity by teeth alone and with the rest of the fossil untouched, he brought witnesses into his chambers before finishing the job. "This operation was done in the presence of people to whom I had announced in advance the result," he wrote, "with the intention of proving to them the justice of our zoological theories, since the true seal of a theory is beyond dispute the ability it has to predict phenomena."

Chiseling the limestone, Cuvier made his way down the animal's body. He found thirteen ribs, and then arm bones that were jointed to allow a creature to climb trees, six lumbar vertebrae in its back—all perfectly consistent with the animal being a relative of opossums. But to satisfy the deepest skeptic, Cuvier set these bones aside. Marsupials, as everyone knew, have pouches, and all of them are supported by two bones that protrude from the front of the animal's pelvis. The pelvis of Cuvier's fossil was facing downward, and so he began to dig into the slab with a sharp steel point, his witnesses doubtless shocked as he destroyed part of the spine of this priceless fossil to get to the animal's front side. He kept digging, tearing out the sacrum and the tail, until at last two hatchet-shaped bones surfaced side by side from the stone in front of the pelvis.

"There is no science that cannot become almost geometrical," Cuvier declared over his pouch bones. "The chemists have proved as much in recent times for theirs, and I hope that the time is not far off when one will be able to say as much of the anatomists."

Almost two centuries later, paleontologists today aren't quite so cocky as Cuvier. An animal's body isn't generated by some formula of its function. It's a product of its evolutionary history, and while correlated progression and the constraints of how embryos develop may give it a certain coherence, there are some fossil animals that are unlike anything alive today and others that seem as if they were stitched together from random parts. But even today teeth remain impressive taxonomic fingerprints, at least for mammals. When we are born we are not yet skeletally set: soft bones have to cure and fuse, to grow stout as we mature, with the exception of teeth. As soon as the cusps and fissures and roots of teeth form in a baby's jaws, they keep their topography for life. If paleontologists find an isolated femur or rib, they often can't say much about its owner, but if they find a bit of cracked, weathered enamel, the only pieces of bone left from a lost mammal dead for dozens of millions of years, they can declare it to be a tiger, a sloth, an opossum, a hominid—and when they do, it is still easy to feel like one of the gentlemen at the museum in Paris, looking over Cuvier's shoulder at the still-sealed slab of limestone and thinking that a magician is at work.

From the beginning, the teeth of whales have helped guide paleontologists to their true history. In 1832, the year that Cuvier died, the president of the Amer-

ican Philosophical Society in Philadelphia received a letter and a box. The letter
was from a Louisiana man who called himself Judge Bry, and inside the box was
a forty-five-pound piece of rock the shape and size of a small powder keg. "A scientific memoir cannot be expected from one who has now spent the last thirty
years of his existence literally in the remotest forests of Louisiana, whose life has
during this long period been entirely devoted to agricultural pursuits," Bry
wrote. Instead, he simply described a hill along the Ouachita River in Caldwell
Parish where he had found fossils of seashells. "About three years ago, after the
occasion of a long spell of rainy weather, a part of the hill slid down near the water's edge and thereby exposed twenty-eight of these bones, which had been until then covered by an incumbent mass of earth about forty feet thick. When
these bones were first seen, they extended in a line, which from what the person
living near this place showed me, comprised a curve measuring upwards of four
hundred feet in length, with intervals which were vacant. The person referred to
destroyed many of the bones by employing them instead as andirons in his fireplace, and I saved what remained from the same fate. If I might presume to express an opinion as to the animal to which these bones belong, I should venture
to say that they were part of a sea monster."

The fossil was put in the hands of a Dr. Richard Harlan. One of the few paleontologists in the United States at the time, he was a homespun version of Sir
Richard Owen. Like Owen, he was a surgeon, working at the Philadelphia Alms
House Hospital, but he was gifted in physiology, geology, zoology, and comparative anatomy as well. He did what science he could when it came to him.
When a sideshow performer arrived in Philadelphia boasting a cure for rattlesnake bites, Harlan tested his claims by running experiments on him. He had
the man's rattlesnake bite puppies, and noted that only one died; he watched
the performer being bitten as well and taking his secret root potion. The man
sweated a bit but survived. Harlan concluded that the snake had little poison in
its fangs. When the remains of animals were brought to the American Philosophical Society, Harlan described them, whether they were condors or orangutans or mammoth teeth. He inspected a skeleton that had been found at the
mouth of the Mississippi and identified as a giant oceangoing lizard, and he realized that it was actually a sperm whale.

Harlan was a careful Cuvierian. Catastrophes had obliterated old residents of
the planet and new ones had taken their place, but "between species and
species, nature has drawn a line of separation which time cannot change, nor
the sophistry of man obliterate," he said. He looked at the giant vertebrae that
Judge Bry had found in Louisiana and, just as Cuvier would do, tried to recon-

struct the whole animal from them. Harlan decided that the vertebrae came from the backbone of some kind of giant, extinct marine reptile, measuring perhaps a hundred feet long. "If future discoveries of the extremities and of the jaws and teeth of this reptile should confirm the indication I have pointed out," he wrote, "we may suppose that the genus to which it belonged will take the name, by acclaim, of BASILOSAURUS"—in other words, the King of Reptiles.

Just after Harlan got Bry's fossil into print in 1835, more bones came out of the south. In Alabama, another judge—this one named John Creagh—found enormous bones embedded in limestone on his plantation, rock so hard he had to blast the fossils free. When Harlan was able to look at Creagh's bones, he realized some of them matched Bry's fossils, and in the same exploded rocks Creagh also managed to find ribs, bones from a giant arm, and jaws with a few teeth still in place. The teeth, each the size of a potato, gave Harlan a little pause. Some of them were narrow, others farther back were blocky. Harlan knew that the teeth of marine reptiles, like those of all reptiles, were uniform, and that only mammals carried different kinds—molar, premolar, incisor—in their mouths. But although *Basilosaurus* had a varied set of teeth, its jaw was long and hollow like the jaws of marine reptiles, and so he stood by his previous claims. The only change he made to them was to guess that his reptile was actually 150 feet long.

When word got to Europe of *Basilosaurus*, it was sucked into a controversy over another set of mammal teeth. Cuvier's opossums in 1804 had come from the earliest part of what is now called the Cenozoic era, from 65 million years ago to the present. As far as anyone could tell, mammals were limited to the Cenozoic; in the preceding Mesozoic era, the only vertebrates were reptiles and fishes. When Lamarckian evolutionists looked at this pattern, they found comfort: it suggested that mammals were only the latest transmutation that vertebrates had gone through. But in 1812 the jaws of a primitive mammal turned up in a piece of slate from Stonesfield, England, dating from the Mesozoic era. The notion that this mammal lived side by side with dinosaurs, unrelated and separately created, was exactly what an evolutionist did not want to hear. The rocks were indisputably Mesozoic, and it was impossible for the evolutionists to deny that the Stonesfield jaws were from mammals because, after all, the teeth were not identical. When Harlan offered the world the reptile *Basilosaurus*, with its varied set of teeth rooted in the same jaw, the evolutionists suddenly had an out: the animals of Stonesfield must be relatives of *Basilosaurus*, and therefore reptiles as well. The Mesozoic period was made mammal-free once again.

The Lamarckians invited Harlan to London in 1839 to address a meeting of the Geological Society. No evolutionist himself, he probably had no idea of the battle into which he was being marshaled, or why Richard Owen was so happy to meet him when he arrived in England. Owen, fresh from his work on the lungfish, the chimpanzee, and the platypus, had been spending his recent years snuffing out the sparks of Lamarck. When he looked at the Stonesfield jaws, he saw mammals, and he was determined to destroy the links that had been used to tie them to reptiles. When Harlan arrived in London, Owen relieved him of the *Basilosaurus* teeth.

A few weeks later at the meeting, Owen took the podium. The features that Harlan thought pointed to a marine reptile Owen brushed away. A long hollow jaw could just as easily be found on a sperm whale. The vertebrae had gentle concave ends like those of mammals, not of reptiles, and the wide hole at their centers—threaded in life by a spinal cord—was the same shape as a whale's. But most important to Owen were its teeth. He had studied them closely, even slicing them open into a series of sections and looking at them under a microscope. While some of the front teeth were simple fangs, each of the rear teeth was a pair of teeth fused together midway, and then fit into a pair of sockets. No fish or reptile known had anything like it; real marine reptiles had teeth with single roots that fused to the jaw itself. Moreover, under a microscope a marine reptile tooth has a simple burst of thin tubes that penetrates regularly spaced concentric rings. But *Basilosaurus* had the same pattern seen in marine mammals, with a thin layer of cement at the surface, and undulating fine tubes just underneath. From his study of the teeth and other bones, Owen was willing to grant that this was a strange beast, but rather than a sea monster, it was a serpent-shaped whale—"one of the most extraordinary of the Mammalia which the revolutions of the globe have blotted out of the number of existing beings."

Harlan went along meekly with Owen's new take on *Basilosaurus*. He returned home to Philadelphia, where in 1842 he learned that Judge Creagh's plantation had regurgitated a sixty-five-foot-long spine with a fair chunk of a head attached, which looked distinctly more like a whale than a reptile. He died the next year, only forty-seven years old, barely known at the time outside the United States and now hardly known within them. But Harlan enjoys a little touch of victory over Owen. In order to completely erase the Lamarckian fiction of reptiles with mammal teeth, Owen wanted to change the name of Harlan's fossil from *Basilosaurus* to the far duller *Zeuglodon*—meaning yoked teeth. "Dr. Harlan," Owen had announced at the London meeting, ". . . has consented that the gigantic monarch, as it was deemed, of the saurian race, should be deposed." But taxonomy is a first-come, first-served science, and the skeleton still bears the name Harlan gave it. Throughout the rest of his life, Owen would have to hear the name of the king that would not be overthrown.

At about the time Harlan was crowning *Basilosaurus* in Philadelphia in 1832, a little museum opened in Saint Louis, Missouri. For a quarter, people could enter and watch magicians and bird imitators; they could look at Egyptian mummies in their sarcophagi, Indian mummies from Kentucky caves, wax figures of Jim Crow and Zip Coon, the famous Siamese twins, live bears, alligators, as well as paintings of the French Revolution, the tunnel under the Thames, and the battle of Austerlitz.

The owner of the museum was a German immigrant named Albert Koch. Like Owen and Harlan he was a paleontologist, but one without schooling or a hospital salary. When he was not running his museum, he was wandering the Mississippi Valley alone looking for fossils, fording rivers, fighting diseases, and eating nothing more than salt pork and bread. In 1839 he dug up the bones of a giant mastodon he named *Missourium* and made it pay for its discovery by taking the skeleton on a tour of America. In Philadelphia, scientists inspecting the skeleton pointed out that he had overblown his beast, adding ten extra vertebrae from a second *Missourium* to its spine and inserting wooden blocks to make it even longer. Rather than let his giant shrink, Koch headed on to Europe to show the fossil off at Piccadilly in 1842, proclaiming it "the sovereign masterpiece and proudest monument of all animated nature." After Owen paid it a visit, the British Museum bought it from Koch for thirteen hundred pounds and squeezed it down to proper size.

Koch knew exactly what to do with the money. He immediately sailed back to the United States, lured by stories of the sea monsters of Alabama. By January 1845 he was in Clarksville, not far from where Judge Creagh had blasted his bones free, and discovered that *Basilosaurus* vertebrae were so common they had become furniture in the region, propping up logs in a fireplace or wedged between beehives. But a full skeleton eluded Koch for months, as he uncovered only shark teeth and other bits of marine bone.

In March a mail rider between Prairie Bluff and Old Washington Courthouse noticed him working on his fossils. He told Koch that he had heard of a ninety-foot-long petrified shark at the courthouse, which people had tried to dig up, only to realize that the bones were too heavy to move. Apparently the mail rider had a fair amount of time on his hands, because when Koch showed interest he rode the forty miles to Old Washington Courthouse to see if the bones were still there. Koch waited nervously for him to return. Three men and a woman from New York stopped to talk to him while the mail rider was gone, and they told him that they were trying to find the sea serpent's fossils as well. Sea serpents were on the public's mind in the United States and Europe, with sailors claiming to have seen living monsters swimming alongside their ships. The possibility that their bones were lying in the wilds of Alabama was irresistible. Two days later the mail rider came back to Clarksville and told Koch that the fossil hadn't been moved.

Koch rode alone to Old Washington Courthouse, across forests wrecked by storms and valleys full of yellow fever. When he reached the little town, he found the fossils that the mail rider had described were not of a shark but still more vertebrae from the sea monster. Koch locked them in the abandoned jail. A white boy and a slave watching him wrestle with the bones told him about fossils still in the ground. They led him to a field by the Sintabogure River, where he saw what he had dreamed of: bones of the great serpent laid out in a huge half-circle.

For three months Koch excavated the Sintabogure fossils. Here around Old Washington Courthouse the bones were even more common than at Clarksville: people used them as garden gates, burned them for lime, set them in chimneys as cornerstones, and slept on them as pillows. Koch hammered at his bones in the scorching sun, burning his skin on the iron tools, working alone after he fired a drunk assistant, fending off gawkers who accidentally broke the giant jaw. When he was through, he had filled five wagonloads with fossils, the biggest haul of whale bones the world had ever seen.

It's hard to judge whether Koch was aware that *Basilosaurus* was actually a whale. While traveling with *Missourium,* he passed through London three years after Harlan's public turnaround. When he found a skull along the Sintabogure, he noted in his journal that it had teeth like a mammal's. If he did know, he certainly kept the truth to himself: by August the New York newspapers were filled with articles about the 114-foot-long sea serpent that was on display in the Apollo Saloon on Broadway, discovered by *Dr.* Albert Koch, as he now called himself. "The serpent of the Deucalion deluge, slain by Apollo Pythius, is beheld, with scarcely the aid of the dullest fancy," said the *New York Dissector.* The *New York Evangelist* was more reverent: "Who knows but he had seen the Ark? Who knows but Noah had seen him from the window? Who knows but he may have visited Ararat? Who knows how many dead and wicked giants of old he had swallowed and fed upon? Perhaps when we touch his ribs, we are touching the residuum of some of Cain's descendants that perished in the deluge."

Scientists were not so kind. As with *Missourium,* they saw that the serpent's spine was not from a single creature. When some of the vertebrae had fossilized they were young and still spongy, while others were old and brittle. The spinal cord canals were slight in some cases, while others were as wide as the braincase itself. The scientists decided that Koch did not have one animal but five, all of whom were not sea serpents but the extinct whale *Basilosaurus.* (In later years paleontologists would recognize that Koch had actually cobbled together several different species of whales.) A gentleman from Old Washington Courthouse wrote to the Boston Society of Natural History, saying how everyone there knew that the bones that Koch claimed came from one animal he had actually found scattered around the Alabama prairie. People there laughed at the solemn reports coming from the New York papers of the great discoveries of Dr. Koch. "Dr. Koch could have made his skeleton three hundred feet long as easily as 114," he wrote.

Koch responded to American critics by fleeing once again with his bones to Europe. He displayed them in Leipzig, and King Wilhelm IV of Prussia was so impressed that he bought the fossils for his anatomy museum, giving Koch a

Albert Koch's sea serpent on display in New York

stipend for life. He went back to Alabama in 1848 and found more bones, which he strung out as a ninety-six-foot-long serpent and brought back to Europe. The new skeleton prompted people to ask Richard Owen if he believed that these giant sea serpents still swam in the oceans, and Owen became so irritated that he wrote a letter to a London newspaper, waving off the accounts from sailors as run-ins with elephant seals and insisting that reptilian giants became extinct with the dinosaurs, while the thing Koch was peddling had been shown by Owen himself to be a whale.

It was the last time that Koch would bother Owen. He settled back down in Saint Louis, calling himself a professor of philosophy. From time to time he would look in Tennessee for seams of iron or coal, but never for a fossil. In 1871 his collection of bones was destroyed in a Chicago fire, but Koch would not be remembered as an absolute hoaxer. While exhuming the *Missourium* he had discovered stone blades around the bones, and he suggested for the first time that Indians had lived alongside the mastodons. Much later he would be shown to be right. He died after a few years of "lingering torpor of the liver," and later, when workmen disinterred his body for reasons unrecorded, they were shocked to find that it had petrified. The paleontologist had become his own fossil.

Before Darwin, taxonomists had a difficult time putting whales in their proper place. Aristotle set them off in a group of their own, as distinct as birds or insects, but by 1700 biologists had recognized that they were actually mammals. When Darwin wrote *On the Origin of Species,* the oldest fossils of whales were the likes of *Basilosaurus,* fully aquatic animals that shared so many of the strange features of living cetaceans that they couldn't show how they might have come from land. After Koch's discoveries, the nineteenth century saw the discovery of a few more ancient whales, emerging mostly in the American South, but also in Kiev, in England, in New Zealand, in Egypt. *Basilosaurus,* it became clear, was only fifty feet long, while other species reached a more modest fifteen feet. In general they were snakier than today's whales, with slender flippers, nostrils close to the ends of their snouts, and a mix of teeth in their mouths unlike the rows of pegs in living odontocetes. Like any whale today, they apparently lacked hind legs and could obviously have lived only in water. How they got there was anyone's guess.

In theory, at least, Darwin didn't think this mystery was uncrackable. "In North America," he wrote, "the black bear was seen by Hearne swimming for hours with widely open mouth, thus catching, like a whale, insects in the water.

Even in so extreme a case as this, if the supply of insects were constant, and if better adapted competitors did not already exist in the country, I can see no difficulty in a race of bears being rendered, by natural selection, more and more aquatic in their structure and habits, with larger and larger mouths, till a creature was produced as monstrous as a whale." The suggestion did not go over well among a public trying to bend their minds around the idea of natural selection. As one newspaper sniffed, "Mr. Darwin has, in his most recent and scientific book on the subject, adopted such nonsensical 'theories'—as that of a bear swimming about a certain time till it grew into a whale, or to that effect." Darwin chose not to fight this particular case of Lamarckian confusion: in later editions of *On the Origin of Species*, he left out most of his speculations about whales.

It was up to the next generation of biologists to think seriously about the origin of whales, and none thought as much about it as the English zoologist William Flower. Flower was twenty-eight when he read *On the Origin of Species*, and he was soon pulled into the bitter debates that followed its publication. His life didn't end up dripping with rancor like those of Owen or Thomas Huxley: his biography reads from start to finish like a pleasant stroll. As a child he built himself little museums in cardboard boxes and then a cupboard, and by the time he had turned twenty-one he was a medical student who had dissected exotic primates such as the galago. He served in the Crimean War, where half his regiment died during the bloody fall of Sebastopol and the battle of Balaclava, but the horror threw no apparent shadow on his later life.

Mammals were Flower's specialty, and when Owen tried to distance humans from apes, Thomas Huxley turned to Flower to investigate. From the audience, Flower was content to watch Huxley storm and attack with the weapons he had fashioned, and only once did he draw much notice. At a scientific meeting in 1862 Owen claimed, as he had many times before, that humans had a structure in their brain called the hippocampus that no other animal (including gorillas) possessed. Huxley said he was wrong and referred the audience to Flower's discovery of a hippocampus in some primates. There the two men were deadlocked until Flower stood from his seat. "I happen to have in my pocket a monkey's brain," he announced. One look and the matter was settled.

Flower had by then taken Owen's old post at the Hunterian Museum, and like his predecessor he dissected his way through a menagerie of animals. He was the first scientist to see the inside of a panda. He examined manatees, chimps, Burmese elephants, pygmy rhinos, sloths, armadillos, howler monkeys, red wolves, and musk deer. But Flower never discovered a new mammal of his own on an expedition and never even nudged a single zoological paradigm. "There is no epoch-making discovery or generalization which can be associated with his name," as his biographer gingerly put it.

Flower had almost as many causes as animals under study—he protested the slaughter of ospreys to decorate hats with their plumes, he raged against the killing of bottlenose dolphins for oil to keep clocks running smoothly. He thought that people should stop the unhealthy practice of burying the dead in sealed coffins and cremate them instead. He weighed in against tobacco, corsets, footbinding, high-heeled shoes, and the fact that the post office charged more to send a heavy letter than a book. Natural selection frightened Owen with its bestiality, while for Darwin, who suffered a series of deaths in his family, God and nature became equally dark. Yet Flower folded evolution peacefully into his complacent Victorian life. He concluded that Caucasians such as himself were biologically superior to the other "non-adaptive" peoples, which would ultimately become extinct. Before they vanished, Flower thought scientists should at least collect their bones for study. Nor did he feel troubled as a Christian evolutionist. Owen's notion of an archetype, an arbitrarily chosen dream of a lampreylike creature, brought no glory to the Creator, but the simple, elegant laws by which life unfolded did. Speaking once to an audience of anxious ministers, Flower asked, "If the succession of small miracles, supposed to regulate the operations of nature, no longer satisfies us, have we not substituted for them one of immeasurable greatness and grandeur?"

Some of Flower's most important mammal work was with whales. He would travel to the coast whenever he heard that one had beached, sketching its corpse in his skeleton notebook and sometimes carting its bones back to London. He showed that the dolphins that lived in the rivers of Brazil, India, and China were all members of the same family. He sorted cetaceans into a new classification, dividing them into the odontocetes and the mysticetes. And toward the close of his career, he gave their origins a lot of thought. Fossil whales like *Basilosaurus*—so aquatic already—offered him little help, but simply comparing living whales to other mammals might be enough, if done in the right way. "Scarcely anywhere in the animal kingdom do we see so many cases of the persistence of rudimentary and apparently useless organisms, those marvellous and suggestive phenomenona which at one time seemed hopeless enigmas, causing despair to those who have tried to unravel their meaning, looked upon as mere will of the wisps, but now eagerly welcome as beacons of true light, casting illuminating beams upon the dark and impenetrable paths through which the organism has traveled on its way to reach the goal of its present existence."

The challenge as Flower saw it was to decide which way the paths went. Were whales the aquatic ancestors of all mammals—a notion of some early Lamarckians—or had their own ancestors lived on land as Darwin had suggested? Mammals as a rule are furry, but the sleek skin of whales is bare save for a few hairs around their blowholes or their chins. Although whales probably couldn't smell, the blowhole still connected to the main airway as did the nostrils of mammals on land. And while Flower assumed that whales must be deaf because the passageway was practically clotted shut—"like a hole made by the prick of a pin"—they still had the ear bones of a mammal. Their fins were actually fingered arms that had been sealed over, and deep inside the blubber at the back of the body were bits of bone and cartilage. In some species, two of these bones fit together like a hip and femur. At the far end of the so-called femur was a capsule of fluid, a vestige of a vanished knee. Even the muscles around these abandoned bones permitted and restrained different movements in the same way that the muscles around a horse's hind legs did.

Flower concluded that some conventional land mammal, complete with four legs, placenta, hair, teeth, and a working nose, gave rise to whales. Ancient whales such as *Basilosaurus* were exactly what we would expect the early aquatic descendants of these creatures to be: they had small skulls, with nostrils closer to the front of their nose. And while living mysticetes had baleen in their mouths and odontocetes had rows of uniform teeth, early whales had a mix of molars and premolars and incisors much like the kind found in the mouths of mammals on land.

Flower cast about for a genealogy leading into the sea. He doubted that seals could be the closest relatives of whales despite the fact that they are intermediate in their adaptation to water. Instead of raising and lowering a giant tail like whales, seals kick their hind legs from side to side, with barely a tail between them. It seemed unlikely that a lineage of animals that had already committed to swimming this way would produce descendants that would shrink their feet back down in order to increase their tails. Flower thought that a whale ancestor might look like one of the big-tailed, semiaquatic mammals like otters and beavers. In a few details in the liver and lungs anatomists had already found similarities between whales and the hoofed mammals (known as ungulates), and so perhaps the names that people had given to whales—sea hogs, sea pigs, herring hogs, and the French *porcpoisson* (from which the word *porpoise* comes)—weren't so far off.

"We may conclude," Flower wrote, "by picturing to ourselves some primitive, generalized, marsh-haunting animals with scanty covering of hair like modern hippopotamuses, but with broad, swimming tails and short limbs, omnivorous in their mode of feeding, probably combining water-plants with mussels, worms, and freshwater crustaceans, gradually becoming more and more adapted to fill the void place ready for them on the aquatic side of the borderland on which they dwelt, and so by degrees being modified into dolphin-like creatures, inhabiting lakes and rivers, and ultimately finding their way into the ocean. Favored by various conditions of temperature and climate, wealth of food supply, almost complete immunity from deadly enemies, and illimitable expanses in which to roam, they have undergone the various modifications to which the cetacean type has now arrived, and gradually attained that colossal magnitude which we have seen was not always an attribute of the animals of this group."

Flower ended with a warning. "Please recollect, however that this is mere speculation, which may or may not be confirmed by subsequent paleontological discovery." The discovery would take over a century, but today paleontologists look back at Flower as almost clairvoyant.

The notion that whales were ocean pigs slipped away. Zoologists didn't dispute that whales descended from some four-legged mammal—if more proof was needed, they could read the report of whalers from Vancouver Island who killed a humpback female in 1919 and found a two-foot-long stub emerging from its flank close to its tail. From the outside, it looked like a baseball bat resting on a hill, and after it was dissected, it was shown to be a series of bones and lumps of cartilage that extended four feet into the body. The cartilage had shrunk by the time zoologists examined it, but they were still able to make out homologies between the pieces and mammal leg bones such as the femur, the tibia, and parts of the ankle and foot. No, the real trouble with Flower's idea was that the earliest whales were clearly killers, with large shearing teeth, while all living ungulates had teeth that they use more like mortars and pestles to eat plants. As the twentieth century matured, more fossils of early whales were discovered. Flower had named them archaeocetes—"first whales"—and they reached back as far as 43 million years. But all archaeocetes were much like *Basilosaurus,* and the three groups of whales—archaeocetes, odontocetes, and mysticetes—had little in common and even less in common with any particular mammal on land.

While the evolutionary dive of whales into the sea remained hidden for most of this century, paleontologists managed in the meantime to sketch out the history of mammals as a whole. When the amniotes first came ashore, they were minor players on the terrestrial stage, but by 320 million years ago the three great lineages of amniotes had arisen that are still alive today, including our own, the synapsids. The earliest synapsids were not particularly distinguishable from other amniotes alive at the time: they were scaly, fanged, hard-of-hearing, cold-blooded, tiny-brained, sprawling things. Yet they already had idiosyncracies that made them our relatives. For example, the synapsids had a single hole in the skull behind each eye, on the borders of which fit jaw muscles that attached to the temple. That same hole persisted in later synapsids, and now it is defined by your cheekbone. Clench your teeth and you can feel the muscles that pass through this hole set up a bulging ridge under your hair. You feel 320 million years of tradition.

We often imagine mammals inheriting the world from the dinosaurs that came before, but that's a parochial view of history. Our synapsid ancestors were actually the earliest amniotes to become successful on land. By 280 million years ago they made up almost three-quarters of the diversity of amniotes. Some branches evolved the ability to eat plants and shared in the great herbivory boom, browsing like scaly cows on shrubs and saplings. Others with teeth like railroad spikes lumbered after other tetrapods, much as Komodo dragons do today. Countless lineages of synapsids rose and fell, now as obscure and exotic as their names: caseids, ophiacodontids, gorgonopsians, venjukoviamorphs. Yet amid the foliage of the synapsid tree, it's possible to make out the gradual, 100-million-year evolution of the rudiments of the mammalian body. And just as the head of lobe-fins evolved in a correlated progression, so did the bodies of these mammal forerunners. Almost nothing happened piecemeal.

Synapsids began to eat in a new way. Rather than swallowing an animal whole or shearing off a mouthful of meat, they altered their jaws, muscles, and teeth so that they could slice flesh in the front of the mouth and chew it in the back. In some lineages they adapted this chewing to eating plants. The benefits were great—the food that hit their stomachs was well ground, allowing their guts to draw more nutrition from it—but to gain them, synapsids had to simultaneously change their breathing pattern. The earliest amniotes had nostrils that opened into the front of the mouth, with the result that they had to hold their breath when they ate. They could afford to since they were only gulping down their food, but synapsid chewing took far longer. As their chewing apparatus evolved, the nasal passage of synapsids also moved to the back of the mouth, with the result that they could both breathe and eat at the same time.

Evolution set them free from other constraints as well. The first synapsids waddled like salamanders or lizards, swinging a sprawled leg forward, planting down a foot, and using the muscles along their flanks to bend into an **S** to increase their stride. Yet the synapsids also needed these same flank muscles to open their lungs and squeeze them shut. They thus had to hold their breath not only when they ate but when they walked. But over millions of years the shoulders and hips of synapsids shifted so that their legs fit below their bodies rather than sprawling out to the sides. Their ribs became rigid, wide-barred cages. Prongs showed up on the sides of their vertebrae, preventing them from bending. As these changes locked out side-to-side movement, synapsids became able to trot, swinging their legs under their bodies and flexing their backs. Now the muscles they used for breathing no longer had to help during walking as well. They were free now to use their lungs as they ran, and with a steady supply of oxygen they could make longer, faster expeditions.

At the same time as their skeletons were shifting, the physiology of synapsids was changing as well, becoming more lively. They could pump more molecules in and out of their cells, burn more food for energy, and build more proteins. And because the chemical reactions that fueled their endurance generated heat as a by-product, their bodies became warm. Growing hair, they trapped the heat as it escaped and became able to colonize cooler climates where they could survive without hibernating. They became able to push into the night. This evolution toward the earliest mammals was thus the product of a feedback between different selective pressures. A creature without a rising metabolism would have no reason to evolve a running gait. At the same time, any small push toward sustained running would have encouraged these metabolic boosts because it would let a synapsid travel farther and find more food.

The earliest synapsids probably heard the world only as a dim hum: strong vibrations might pass into the skull, but the stapes, still helping to hold the braincase to the rest of the head, could not respond. In time, the synapsid skull cemented itself, and sounds that might hit the back of its jaw could now make the stapes vibrate. The ear became more sensitive as synapsids evolved the ability to hear through their jaws. An early synapsid jaw was a committee of bones: the teeth lodged in a large bone called the dentary, which fused at the rear to a set of smaller bones that made contact with the skull. As synapsids evolved a more powerful bite, the dentary expanded while the other bones shrank and loosened. As they were rearranged, these extra jaw bones—still essential for opening a synapsid's mouth—now came into contact with the stapes. Any vibrations that might hit them traveled into the ear.

Most synapsids were husky daytime bruisers, but one ratty little lineage around 240 million years ago specialized in nocturnal hunting. As they explored the world beyond twilight, they needed to squeeze more information out of their eyes, noses, and ears, and to do so, they needed bigger brains. When they grew as embryos their brains swelled faster and longer than before, and much of the extra tissue became a neocortex, an outer husk of brain dedicated to taking in and integrating the information from its senses. The swelling—a straightforward change in developmental timing championed by Stephen Jay Gould—produced one of Guenter Wagner's beloved developmental collisions. The ballooning brain pushed out to the position of the loose rear bones of the jaw and then kept pushing, tearing away the bones and carrying them away. The jaw, now made only of the dentary, formed a new joint of its own with the skull. The kidnapped rear bones could now do nothing but listen, and they dwindled into the delicate incus and malleus of the mammalian middle ear, small enough to vibrate in response to high-frequency sounds. These new signals entered a brain that was now awash with signals. They might have overwhelmed an animal with a smaller brain, but the early mammals had enough processing power to organize them into representations of the world. No longer did they rely on the reflexive responses of their ancestors. For the first time there were objects and labels. There were minds.

As impressive as all this evolution might seem, the earliest mammals watched the synapsid dynasty that had created them crash, with most of the lineages becoming extinct. By 220 million years ago another group of amniotes had originated, having changed from sprawling reptiles to upright runners. They began initially as small, two-legged carnivores, but quickly they evolved into forms ranging from four-legged, long-necked plant-eaters to forty-foot-long carnivores. These were the dinosaurs, and they dominated the land for 150 million years. In the ocean, relatives of lizards—long-necked plesiosaurs, swordfish-shaped ichthyosaurs, and others—grew up to eighty feet long. Close relatives of dinosaurs known as pterosaurs grew to the size of small airplanes and glided over deserts or skimmed fish from the oceans.

Paleontologists still debate exactly why descendants of *Tyrannosaurus, Iguanodon, Triceratops,* and other big dinosaurs still aren't holding mammals at bay. By some measures, many dinosaur lineages were failing toward the close of the Cretaceous period, which ended 65 million years ago. The climate cooled, seas retreated, and perhaps dinosaurs could not adapt to the changes. (By contrast, crocodiles, lizards, snakes, birds, and most fishes and sharks passed over the boundary with few losses.) It's almost certain, though, that the Cretaceous period ended with a cosmic exclamation point: an asteroid or comet measuring ten miles across dropped onto the Yucatan Peninsula, vaulting sulfates and carbon dioxide into the air and possibly setting much of North America ablaze. On the other side of the planet at about the same time, volcanoes in India disgorged enough lava to cover Alaska and Texas in a quarter-mile-thick pavement and added their own pernicious cologne of poisonous gases to the atmosphere. While some groups of organisms may have been declining for millions of years, the end of the Cretaceous did bring a violent, swift end for many life forms, particularly in the ocean.

Some mammals survived the extinctions, and by this time they were very different from the founders of their class. They could nurture their young with milk, which meant that their offspring had much better chances of surviving to maturity. Some mammals—the platypus and its relatives—still laid eggs, but in newer lineages the embryo and its membranes remained in its mother's womb. Among marsupials, the embryo crawled into a pouch, while among placental mammals, the baby stayed in utero until it was ready to emerge. The modern orders of placental mammals had only just begun to split when the unfeathered dinosaurs died, the earliest primates not looking much different from the tree shrews, the grandfathers of elephants barely distinguished from the grandmothers of horses. But mammals were bursting forward into new forms at an extraordinary rate, and it was during this bustling time that something somewhere became a whale.

For much of this century most paleontologists guessed that whales descended from some carnivorous mammal on land. The first hint that this wasn't the case didn't come from their bones but from their cells. Just as mutations in genes can resculpt bones, they also change the structure of proteins. A mutation that leaves an energy-generating protein in shreds may kill its owner while a different one might make the protein more efficient, but many mutations have no effect on proteins at all because much of their structure isn't vital to the tasks they do. An altered amino acid here or there goes unnoticed, and as a population of animals begins to cleave from the rest of its species, these mutations build up, forming a pattern in their proteins distinctive to themselves.

In 1950 biologists at Rutgers University used proteins to find a place among the mammals for whales. They extracted a blood protein from fin whales and pygmy sperm whales, as well as from a wide range of other mammals such as coyotes, cows, pangolins, elephants, kudu, hedgehogs, and warthogs. They injected a little of the whale protein into rabbits, whose immune systems responded by scrabbling together antibodies that could clasp on to these foreign molecules and notify killer cells to attack. The biologists then drew off the rabbits' blood cells, leaving behind a clear serum still carrying these whale antibodies. If they added the whale protein to this serum, the antibodies would wrap themselves around them and the reaction would make the fluid turn cloudy. Adding the proteins of other mammals also provoked a response, albeit a gentler one. The antibodies, tailored especially for the whale protein, could attach themselves with mixed success to those from other species if they had a similar structure—the more similar, the better the lock. Proteins from most of the mammals drew barely a notice from the whale-primed rabbit antibodies, the researchers found, but cows clouded the serum. That demonstrated that the proteins of these ungulates were probably more whalelike than the proteins of any of the other mammals. As far as the rabbits were concerned, Flower was right.

Yet any comparison of living animals, no matter how sophisticated, could never tell the full history of whales because too many intermediates have disappeared. It had to wait for paleontologists to take a Cuvierian interest in whale teeth. In 1966 Leigh Van Valen, then a graduate student working at the American Museum of Natural History, was overhauling the taxonomy of carnivorous mammals of the two periods that followed the extinction of the dinosaurs, known as the Paleocene and Eocene. Paleontologists had traditionally put many of these mammals into two orders, the creodonts and condylarths. The creodonts were the clawed predators of the time, and paleontologists accepted that true carnivores—cats and dogs and other predators alive today—descended from some unknown member of the order. Condylarths included the hoofed forerunners of today's ungulates—including the even-toed artiodactyls such as pigs and cows, and the odd-toed perissodactyls, such as rhinos and horses. Like their descendants, condylarths were plant-eaters of various shapes and sizes. But while it's easy to tell the difference between the skeleton of an elephant and a lion today, these earlier mammals were far less specialized and far more confusing. The truth was that *creodont* and *condylarth* were names glued onto wastebaskets into which paleontologists tossed fossils that looked generally similar but were hard to classify.

Van Valen was one of the first paleontologists to root seriously through the trash. One of his more remarkable acts was to find a taxonomic home for an extinct group of mammals known as mesonychids. Paleontologists had put mesonychids among the meat-eating creodonts because of their massive shearing teeth. But even in a category as baggy as the creodonts, mesonychids were oddball because their legs looked more like an ungulate's hooves than clawed paws. Their real identity, Van Valen argued, was in their teeth, which looked like those of condylarths. Condylarth teeth are complicated affairs. Their first upper molar, for instance, is dominated by a triangle of cusps connected by ridges, with extra cusps scattered along the edges, all surrounded in turn with outer rings of raised bumps. This kind of tooth does well for catching tough food such as leaves or grass in numerous little clamps and slowly pulling it apart with jaw movements. Many living ungulates—descended from mammals with this kind of tooth—have taken the architecture to its evolutionary extreme: all of their teeth now look like molars.

Mesonychids, Van Valen argued, were also descendants of plant-eating condylarths, but they took an evolutionary route in the opposite direction from living ungulates. Think of teeth as furniture: condylarths had plain Mission style, and living ungulates have evolved them into Rococco. Mesonychids went Shaker. In their first upper molar, for example, they retained a triangle of cusps but stripped away the ridges, bumpy rings, and extra cusps. No longer could they grind food with these teeth; they had been simplified to cut flesh. If you wanted to transform a goat so that it could hunt for its dinner, this is how you would change its teeth.

At the same time Van Valen had been looking over the teeth of the earliest whales known at the time, and he noticed that they were surprisingly like those of a mesonychid. If you raised up the outward cusps of a mesonychid molar and sandpapered down the inner cusp, it would fit into an archaeocete's mouth. Van Valen also pointed out some of the holes in the back of the skulls of mesonychids and whales that were similar, but it was the teeth—those dental fingerprints—that made people think that he might be onto something. From a distance, a mesonychid might seem to have as much in common with a whale as a Dodge does with a submarine, but nevertheless, mesonychids became the best candidate—the only, actually—for the whales' closest relatives on land.

Van Valen, now a professor at the University of Chicago, has been gone from the American Museum of Natural History for thirty years, but from time to time young paleontologists return to study the museum's fine collection of mesonychids from Mongolia and Wyoming. One of them is a woman named

Maureen O'Leary. I paid her a visit, and we walked through the Hall of Fossil Mammals. When Van Valen was working at the museum the hall was as gloomy as a lost memory, but now windows have been knocked through, the pale sky over Central Park lighting up its walls. Children stopped to smack the steel-plated computers before running on to the dinosaur hall. Television screens showed running gazelles, and humpback whales squeaked out their songs on loudspeakers. We stopped in front of a glassed-in display of mesonychids. As O'Leary pointed to skull crests and tooth angles, her deep voice rattled with a cough she'd been fighting off for six months. It probably had something to do with the formaldehyde that preserved the human cadavers she used to teach medical students at New York University their gross anatomy—a common job for paleontologists in these lean academic times. "It's fun cutting into people, but there are a lot of respiratory insults in my world," she said.

In the mesonychid display was one of Charles Knight's handsome paintings from the turn of the century. "That's *Mesonyx*," O'Leary said. *Mesonyx* lived in Wyoming during the Eocene, and Knight painted it roaring over a corpse it was scavenging. The only thing odd about the picture was how the animal's back legs flopped to one side like a basset hound's. "I don't know why it has that strange view—it looks like it's dragging its hind legs," said O'Leary. Maybe Knight was trying to get across the idea of a scavenger and wanted to give it the undersized back half of a hyena—a popular analogy for mesonychids. O'Leary wasn't so sure. "I don't know," she said. "It seems rather ad hoc to me."

O'Leary's doubts are nothing new in the study of mesonychids. They do not want to fit into the categories of living animals. One of the first descriptions of a mesonychid was by Edward Cope, who in 1884 did his best to apply the English language to a beast now called *Pachyaena* (a description that later turned out to be utterly wrong):

> *The forelimbs are so much shorter than the hindlimbs that the animal customarily sat on its haunches when on land. In walking, its high rump and low withers would give it somewhat the figure of a huge rabbit. Its neck was about as long as an average dog. Its tread was plantigrade [its soles rested flat on the ground], and its claws like those of various rodents, intermediate between hoofs and claws. The animal, to judge from its otter-like humerus, was a good swimmer, although there is nothing specially adapted for aquatic life in the other bones of its limbs. Its teeth, on the other hand, are of the simple construction of mammals which have a diet largely composed of fishes. We*

cannot but consider this animal as one of the most singular mammals which the Eocene period possessed.

O'Leary led me away from the public halls to the museum's hidden mazes of storage and offices. Her office was a table lost among the museum's cabinets of mammal fossils. A cherry-colored cast of a molar from a newly discovered mesonychid sat on her computer monitor. Cuvier would not have been happy studying mesonychids: despite the similarity of their teeth, it's hard to understand what mesonychids were because they were so different from one another. To illustrate this point, O'Leary pulled open a cabinet drawer next to her desk and brought out lower jaws the size of sugar spoons. They belonged to an animal called *Haplodectes,* and as tiny as it might have been, its teeth are a close match to the more than twenty larger mesonychid species.

To give me a sense of how big mesonychids could get, O'Leary then walked over to an open shelf and leaned over a bowl made of cardboard and aluminum foil. Inside was a cranium resting upside down, so enormous that it would have been dangerous for the two of us to try to take it out and turn it right side up. Inspecting it from a palate perspective, I could see how I could slide my arm elbow-deep into the long mouth of this animal, and how easily it could snap it off with its huge teeth. No one has found another bone from the skeleton of this terror, named *Andrewsarchus,* but some paleontologists have suggested that it may have measured twelve feet long, or two lions put head to hip. If so, it would have been the biggest meat-eating mammal to ever live. Look at its teeth, and you see similar, although oversized, versions of the ones belonging to the rat-sized *Haplodectes.*

Such a wild variety makes it hard for researchers like O'Leary to get mesonychids to sit obediently on the branches of a cladogram. They agree that mesonychids are ungulates, but they haven't decided yet whether they are closer to the artiodactyls or the perissodactyls. Nor have they come to an agreement on how mesonychids are related to each other. They may not even form a natural group, all descending from a common ancestor. And most unclear of all is which mesonychid was the closest relative to whales. Nominations for cousin to cetaceans have run the gamut, from little *Haplodectes* to bear-sized *Pachyaena* to monstrous *Andrewsarchus.*

As she earned her Ph.D. under Ken Rose at Johns Hopkins University, O'Leary made a special study of *Pachyaena.* Using the shapes and proportions of its bones, she has tried to come up with a profile of it as a living creature, and while Cope's shambling rabbit-bear is clearly a fantasy, it's been hard for her to find modern analogues to take its place. "They seem to be this chimera of different things," says O'Leary. "Sometimes they leave you stranded."

Mesonychids stood on hooves—flattened elongated toes—that looked much like those of a pig or a tapir and gave them a stable base for running. The geometry of their gently curved leg bones matches those of runners such as dogs and wild boars. All of their joints speak of pure dedication to racing, as opposed to digging, climbing, or anything else mammals do with their limbs. Their ankles and wrists swing only in the running plane, and their elbows and knees are locked in as well. Where their legs fit into the shoulders and hips, there are crests and flanges to keep them from dislocating at high speeds. "They were maybe not as fast as some modern runners like cheetahs or gazelles, but they were like tapirs, who achieve a decent speed when they get going," says O'Leary.

But their speed doesn't make sense to O'Leary. "Why did they run? Did they run to catch something? If they did, it doesn't look like they'd be able to catch it." A mesonychid could hardly sink its hooves into a condylarth, nor did it have the quick, killing bite of many predatory mammals. True carnivores all have perfected a set of dental scissors made out of their fourth lower premolar and first upper molar, and using the blades for years on end wears them down in a distinctive pattern. Mesonychids couldn't slice this way, suggesting to O'Leary that they did only a half-decent job of cutting flesh. Yet their skulls did have a giant crest running along the top, where jaw muscles could anchor and generate a powerful bite. Mesonychids had tall shafts on the vertebrae in their necks

The mesonychid *Pachyaena*

to hold up their heavy heads and give them power that they might have used as they wrenched and tore at flesh.

This is the point at which paleontologists often turn to the hyena for inspiration. Hyenas, which are true carnivores closely related to cats, have wide, powerful skulls and a rigid backbone (which resembles that of a mesonychid) that is suited for steady runs over long distances. Like hyenas, mesonychids might have trotted far across the Eocene plains feeding on the carcasses of other condylarths. If their teeth weren't sharp enough for the hunt, they could at least use them for the cleaning up. But O'Leary isn't satisfied: "The hyena analogy has been batted around a lot, but I'm not sure how good an analogy that might be as scavengers. Hyenas do a lot of their own killing. Would it be possible to survive purely on corpses, or would you have to eat something else as well?"

While the teeth of mesonychids aren't worn like a lion's, O'Leary has always been struck by how badly beaten up they tend to be. "You can see right through to the dentin, which suggests they were chewing on something really hard," she explains. Mesonychids might have roamed along rivers or coasts, eating animals like fish and turtles, while never turning down a corpse. Growing up along tropical rivers, mesonychids might have had to face lunging crocodiles, and in such a case, O'Leary and Rose have suggested that the juveniles could have used their fast legs to get away from an attack. And perhaps along those rivers, the snout of one lineage of mesonychids lengthened, making it easier to catch fish, and gradually they became comfortable in the water. They might have taken the first steps toward evolving into whales.

When O'Leary and I had finished talking, I walked down the wide museum steps to the subway station, and there I imagined the deserted platform filled with mesonychids, little *Haplodectes* skittering like a rat on hooves past my feet, *Andrewsarchus* looming by the token booth and roaring as it swung its head angrily at a congregation of *Pachyaena,* sending them trotting in my direction. I had a hard time magnifying any of them to the size of a whale, smoothing and blubbering over their bodies and putting their nose above their eyes. But then I had to remember that I myself would do a pretty bad impression of a lungfish In hindsight macroevolution may look obvious or destined, but in the present its future is impossible to know.

Chapter Seven

ALONG THE TETHYAN SHORES

Like most paleontologists, Phillip Gingerich has an office that looks more like a giant supply closet. Even though the back corridors of the University of Michigan's Museum of Paleontology are hemmed by high wooden cabinets of fossils, Gingerich keeps hundreds more in his own office, laid out in steel drawers, resting on yellow foam and papers full of Ann Arbor news. He fills most of the remaining space with walls of books and filing cabinets full of papers. There may be ten thousand papers shimmed in their drawers; he can't say. The office is made of two rooms; one is windowless, and the other has its black blinds drawn. They both smell of Murphy's soap. I had been deposited in his office one morning, early for an appointment, and I was surprised that it didn't bear much of an aura of this paleontologist who, over the past twenty years, has found some of the most remarkable whale fossils in the world. But gradually I noticed telling things in the unclaimed spaces: the postcards from Pakistan on the wall, the dolphin's skull tucked in one corner of a shelf, a box hanging close to the ceiling that contained a dog skull above crossbones of humeri and the ancient saying *De Mortuis Nil Nisi Bonum*—do not speak ill of the dead. And taped to Gingerich's inner office door is a sheet of paper with little photocopied Rockwell Kent illustrations of whales and ships and a few paragraphs a friend once wrote for him as a going-away-on-an-expedition present. It begins, "Call me Gingerich."

What follows is a patching-together, for the most part faithful, of the beginning of *Moby-Dick*. The one change is of Ishmael's destination of Cape Horn and the Pacific: "The great flood-gates of the wonder-world swung open, and in the wild conceits that swayed me to my purpose, two and two there floated into my inmost soul, endless processions of the whale, and, midmost of them all, one grand hooded phantom, like a snow hill in the air. I stuffed a shirt or two into my old carpet-bag, tucked it under my arm, and started for Cairo and the Mediterranean."

Below it was another passage from a later chapter. It needed no personalizing to fit Gingerich:

> *Ere entering upon the subject of Fossil Whales, I present my credentials as a geologist, by stating that in my miscellaneous time I have been a stone-mason, and also a great digger of ditches, canals, and wells, wine-vault, cellars, and cisterns of all sorts. Likewise, by way of preliminary, I desire to remind the reader, that while in the earlier geological strata there are found the fossils of monsters now almost completely extinct; the subsequent relics discovered in what are called the Tertiary formations seem the connecting, or at any rate intercepted links, between the antechronical creatures, and those whose remote posterity are said to have entered the Ark; all the Fossil Whales hitherto discovered belong to the Tertiary period, which is the last preceding the superficial formations. And though none of them precisely answer to any known species of the present time, they are yet sufficiently akin to them in general respects, to justify their taking rank as Cetacean fossils. But by far the most wonderful of all cetecean relics was the almost complete vast skeleton of an extinct monster, found in the year 1842 on the plantation of Judge Creagh, in Alabama. The awe-stricken credulous slaves in the vicinity took it for the bones of one of the fallen angels. The Alabama doctors declared it a huge reptile, and bestowed upon it the name of* Basilosaurus. *But some specimen bones of it being taken across the sea to Owen, the English Anatomist, it turned out that this alleged reptile was a whale, though of a departed species. So Owen rechristened the monster* Zeuglodon; *and pronounced it one of the most extraordinary creatures which the mutations of the globe have blotted out of existence.*
>
> *When I stand among these mighty Leviathan skeletons, skulls, tusks, jaws, ribs and vertebrae, all characterized by partial resemblances to the existing breeds of sea-monsters; but at the same time bearing on the other hand simi-*

*lar affinities to the annihilated antechronical Leviathans, their incalculable
seniors; I am, by a flood, borne back to that wondrous period ere time itself
can be said to have begun; for time began with man. Then the whole world
was the whale's; and, king of creation, he left his wake along the present lines
of the Andes and the Himmalehs.*

Before I met Gingerich I had imagined that he had been set on chasing down
fossil whales since he was a boy, but his love of antechronical leviathans actually
came late. "I grew up in the Midwest," he told me after he arrived. "To me, a
whale was something from another planet." Gingerich is a tall man with a
smooth, calm face, who talks about intricacies of statistical sampling and evolutionary theory at a farmer's unhurried pace. When he came east to study at
Princeton in 1964, he was thinking about becoming an economist. "Like most
people, I took geology in my first term because it seemed like the easiest way to
satisfy the science requirement," he says. By the second term he was taking advanced geology classes, in the summer he was hammering rocks in Montana,
and two summers later he was digging out 55-million-year-old jaws of the primates that once leaped through the jungles of Wyoming. "You don't take jaws
full of shiny teeth out of the ground very long before you're hooked."

After teaching science in Malawi for two years, Gingerich went to graduate
school for vertebrate paleontology. He dug out more mammals in Wyoming
during the summers, still mainly primates, but also ancient artiodactyls, rodents, and mesonychids. One of the many questions he mulled as he worked
with them was how and when these modern orders had arisen. (When paleontologists say "modern" they can, as in this case, mean 50 million years old.) "We
pretty quickly established that the modern orders we were interested in come in
at one bed," says Gingerich, "and we can take you to Wyoming and you can put
your hand on that bed; below it you will not find them and above they quickly
become common. If everything comes in together, you start to think it had to
evolve somewhere else."

Out of ignorance came many possibilities. The North American mammals
could have evolved in Central America and come north, in Europe and crossed
west through Greenland, in Asia and swept east through Alaska. The last possibility seemed like a particularly good one to Gingerich, but politics hid Asian
fossils in places such as the Soviet Union, China, and Mongolia as effectively as
a swamp. Pakistan, however, was willing to take in a curious American, and
Gingerich had read accounts of how oil geologists in the 1930s had discovered

a few bones of Eocene mammals near the village of Ganda Kas in the bare foothills of the Kala Chitta mountains. A British paleontologist named Guy Pilgrim studied the fossils and concluded that some of them were the teeth of mesonychids. "Pilgrim described some of them as teeth of mesonychids from Pakistan, and we find mesonychids in Wyoming, so I thought it would be good to follow up on that. The best place to find another fossil is where you've already found one."

In November 1975 Gingerich took two weeks off from a tour of primate jaws in the museums of Europe to follow Pilgrim's trail through Pakistan. He traveled to the town of Rāwalpindi, and every other morning he hired a taxi to take him fifty miles west toward the Indus River. There the Kuldana and Kohat formations, beds of Eocene rock in the neighborhood of 50 million years old, peeked through the scrubby hills. The rocks had formed as the Indian subcontinent rammed into Asia, forming the Himalayas. In so doing, India closed off the eastern end of the Tethys Sea, a body of water that once reached from Spain to Indonesia. Later the Middle East and North Africa would also move so far north that the Tethys was reduced to what is now the Mediterranean Sea. India's collision wildly raised and dropped the sea level on the eastern shores of the Tethys, and Gingerich could see that the rocks of Pakistan—which formed part of the leading northern edge of the Indian subcontinent—bore witness. The low outcrops around Ganda Kas were striped with chocolate mudstones and limestones the color of old dollar bills—switching from dry land to rivers to bays and back in a few hundred yards. There Gingerich would walk for miles looking at the surface float—the loose rocks that had fallen from the outcrops. When he spotted bits of exposed bone, he stopped to put them in his backpack. In the afternoon he would trudge back to the road, the cab driver would wake from his nap under an acacia tree, and they would head back for Rāwalpindi.

The rocks were not spectacular, and he gave the paper in which he described them a suitably sleepy title: "A small collection of fossil vertebrates from the Middle Eocene Kuldana and Kohat formations of Punjab (Pakistan)." There were some crocodile bones and fossilized scat, crab claws and shark teeth. The mammals left him only bits of their teeth, and he identified them as anthracobunids (relatives of elephants and manatees) and dichobunids (early artiodactyls the size of squirrels). Most of the teeth of mammals came from rocks that were clearly formed on land, but Gingerich also found a tooth from a mesonychid called *Gandakasia* in a stretch of limestone that also had clams and other marine fossils. Gingerich knew about Van Valen's link from mesonychids to whales, and so he wondered if *Gandakasia* was actually an early whale. It was not a thought he pondered much; he was dedicated to finding land mammals, and any rock that held a whale was not going to be interesting to him.

The other fossils made Gingerich want to come back to Pakistan, to look for more counterparts to his Wyoming herds. But two months after his visit a separate team of Pakistani and American paleontologists had come digging around Ganda Kas and found fossils of their own. When they all realized that they were working at the same place, they met to settle the claims. Gingerich lost. "We decided that they would keep working in the central, plum area only three hours from town. I was supposed to work on the west side of the Indus," Gingerich explained. He didn't sound resentful as he recalled the decision, but then twenty years had passed. When he came back to Pakistan in 1977 with a team of paleontologists from Michigan and Paris, they headed for new territory: the Rakhi Nala Valley at the base of the Sulaimān Range, where the green Eocene shales of the Domanda Formation were bared on the valley walls.

Pilgrim had also found a few bones here, a place even more remote than Ganda Kas. A few farmers raised millet among hundreds of miles of empty hills crossed by a single road that was clogged with trucks bringing fruit down from Baluchistan. As the paleontologists on Gingerich's team found a few pieces of mammal bone here, the sedimentologists studied the shales to divine the landscape where they had formed. Gingerich was not happy when they told him what they had found. "All marine," Gingerich remembers, the despair still in his voice. The mammal bones they had found had not been laid down on dry land but had been carried out to sea. "After a week we put our tail between our legs and fled to the backup option in the south of Pakistan, where we hoped we might find Paleocene fossils. After a week or two, we found a couple pieces of turtle that were perfectly boring. Then we fled back north to option three. I don't remember much there. I got sick and I remember the fever starting in the afternoon, me lying under an acacia tree, and a Pakistani colleague finding a little mammal jaw."

That jaw was the sign at last of terrestrial mammal fossils. Gingerich's group had found it only thirty miles west of Ganda Kas, near a village called Chorlakki. They came back the next year and again in 1979 and harvested the fossils. "The fossils at Chorlakki are all coming out of a layer as hard as cement in a red bed sequence," says Gingerich. "Little artiodactyl jaws and mammal teeth run all through it, so we got to the point where we were just breaking the rock up. If there was anything promising we were taking the matrix back and processing it to get small mammals out of it." Meanwhile Neil Wells, a University of Michigan graduate student, was trying to figure out what sort of landscape these mammals were living in. He explained to Gingerich how they were standing near what was once a strand by the shores of the Tethys almost 50 million years ago, where freshwater streams full of fish cut through flat scrub. The streams were usually only a few inches deep, but when seasonal rains flooded them they would suddenly lurch over dry land, occasionally drowning any mammals that might be in a wooded hollow nearby, such as grazing artiodactyls, primates, mesonychids, bats, and shrewlike insectivores.

One of Gingerich's fellow rock smashers, a French rodent expert named Jean-Louis Hartenberger, brought his hammer down one day on a rock that broke open to reveal a three-inch slot of bone. Hartenberger was annoyed that it was obviously not part of a rodent's skeleton, but to see it more clearly, he thwacked the rock again. This time the top of a skull was revealed. "All that we could see was that it was a skull, and it had no brain," says Gingerich. A huge crest rose over its head like a mohawk of bone, leaving little room for a braincase to hold its gray matter.

Gingerich brought the skull home with the other fossils, and the preparators at his lab worked on it for a year, knocking away bits of the cement-hard rock with an air-scribe. They exposed the back of a skull of a coyote-sized animal. On top of the skull was the massive sagittal crest that Hartenberger had first spotted in Pakistan, but hanging below the skull was something that Hartenberger hadn't seen: a bone the size and shape of a grape.

It looked like the shell in which living whales keep their ear bones sequestered so that they can hear underwater. Gingerich inspected the edge of the shell where it attached to the skull, and he saw an S-shaped flourish of bone serving as an anchor. Every whale, alive or extinct, has this little signature of bone, to the exclusion of all other animals on earth. "So there I was with a whale and knew nothing about them."

Working with his colleague Donald Russell of the National Museum of Natural History in Paris, Gingerich named the fragment *Pakicetus* and began to educate himself on fossil whales. The ears were distinctly cetacean but nevertheless peculiar. Whales can hear directionally underwater because they keep their ears in a bony casing separate from their skull, floating in a loose basket of ligaments and foam, so that they receive sound only through their jaws. The grapelike bones of *Pakicetus*, by contrast, were still glued firmly to the bottom of its skull at three different places. It might have been able to hear reasonably well underwater, but vibrations could still seep through the bony attachments, blurring the direction of the sound. When living whales dive, they fill a network of spaces around their ears with blood to continue keeping the ears isolated at high pressures. *Pakicetus*'s skull was solid in the corresponding area, suggesting that it couldn't swim deeply. Here was a whale not yet comfortable underwater, unable to hear or swim well, found not in the ocean but in a shallow stream—both the most primitive whale known and the oldest by several million years.

Once Gingerich would have groaned to have found a whale fossil, but *Pakicetus* was too remarkable to ignore. He went back to Pakistan no longer to discover where his Wyoming mammals had come from but to find more bones of *Pakicetus*. He worked again around Chorlakki, but he found only a few of the whale's teeth. They were certainly important—they were even more like mesonychid teeth than the ones that had inspired Leigh Van Valen to connect them to whales—but Gingerich could find nothing from the neck down. In the meantime civil war had exploded in Afghanistan, with millions of refugees making their way into northern Pakistan. The government didn't like the idea of foreigners crawling around the back country and adding to the chaos, and when Gingerich applied to return in late 1982 he was refused.

But fossil whales had become part of Gingerich's life, and he managed to find another way to study them. A primatologist called him, having just spent a field season in Egypt at a camp run by Gingerich's old teacher at Yale, Elwyn Simons. Each year Simons went to the deserts west of the Nile to dig in Oligocene rock, looking for primates close to the common ancestor of monkeys and apes. Sometimes a few of Simons's crew grew restless and drove deeper into the desert for a spell. Fossil whales had been found in rocks a few hours away from their camp

The head of *Pakicetus*, the oldest known whale

in 1906, and some people in Simons's crew decided to follow up one of the reports. Apparently they had discovered a fossil lying on the desert floor like a giant sand snake.

This fossil whale was a 40-million-year-old specimen of *Basilosaurus*, and Gingerich decided it would be a good thing to have. He could use it in his work with *Pakicetus*, to see how whale ears evolved during their first 10 million years. In the fall of 1983 Gingerich headed for Cairo, although not alone like Ishmael; with him went his wife, anthropologist Holly Smith, and Michigan paleontologist Gregg Gunnell.

A century earlier to the year, a German paleontologist had found the first Egyptian whale fossil. He recorded it as being located in the ♀hills ("How does one refer to such unnamed hills?" he asked himself.) He would try his best, noting how many camel-hours they were from the nearest mosque or abandoned temple. Other paleontologists came back every decade or two to the Eocene rocks that run west of the Nile, and they stumbled across new whales at sites that later paleontologists could never manage to locate. Although a few bones of one 45-million-year-old whale had been found in these rocks, most were of a *Basilosaurus* vintage. In 1947 a University of California expedition heading across Africa came to the desert in four U.S. Army trucks. The scientists stopped their caravan and walked for miles across the sand, scanning for whales. "Very often a fragment of bone about one centimeter square or less was the first indication of the existence of fossils in an area that appeared blank," wrote P. E. P. Deraniyagala, a Ceylonese scientist who was part of the expedition. "Searching in the vicinity presently revealed other fragments which became more numerous as one approached the fossil. Following up the trail of fragments resembled tracking a wounded animal by its blood spoor, only here one did not know whether the prize would be something extremely valuable or utterly worthless. It is this element of chance that keeps the fossil hunter working at fever pitch all day long in spite of the terrible blinding white glare of the desert and its aridity."

On their first visit, Gingerich and his team drove four hours south from Simons's camp along the ridge of Oligocene rock and plunged their jeeps down a slope of dunes onto an Eocene plain called Wadi Hitan. This was the place that Simons's students had visited; it had first been discovered by British scientists who had found so many fossil whales that they had named it Zeuglodon Valley. With only camels to bring back the bones, the British had left most in place. Gingerich spent two days digging out the *Basilosaurus* that Simons's students had seen, while Smith walked the valley, spotting whale skeletons all around. Zeuglodon Valley might not be the birthplace of whales, but what it lacked in primordiality it made up in numbers. In 1985 Gingerich returned, this time setting up a little camp of three tents in the valley so that his crew could stay among the whales for a week at a stretch, driving back to Simons's camp to restock and take sponge baths. The only company they had in those months were vast domes of sand and a carpet of bones. Once a few camels quietly walked by the camp, their hooves inscribing the sand. Three days later a Bedouin rode by in pursuit. He was the one human they saw.

Fossils overwhelmed them: almost every day they found a new whale, and the total would ultimately reach 349. "It was quickly evident we weren't going to bring many of them back," says Gingerich. "Instead we started mapping them. Some are skeletons that are just coming out and others are being blasted away. You get them where they're just peeking out, and that's what we like, because then the whole skeleton is in there and well preserved." Sometimes, though, winds had scoured away the sand and exposed the fossils entirely. "They're out there like big giant snakes. I wonder why there aren't any references to this place in ancient Egyptian sources. It would have been accessible to them, and I wonder . . . I wonder if their crocodile worship wasn't partly based on *Basilosaurus* skeletons."

Gingerich had now come to the western end of the Tethys Sea. *Pakicetus* had lived 49 million years ago on the eastern edge of the sea, at the shallow end where it was saltiest, but the whales that lived in Zeuglodon Valley about 10 million years later swam where the Tethys streamed toward the Atlantic. In Zeuglodon Valley are braided worms of rock that were once mangroves; they grew on the edge of a shallow bay full of giants—sharks, sea turtles, sea cows, and crocodiles that swam with the whales. It is hard to guess why so many whales should be found in so small a place. Deraniyagala had suggested that they were a fossilized school of beached whales, but the animals had actually died in the bay over thousands of years. Many of the whales were a fifteen-footlong species called *Dorudon atrox*, and Gingerich's teams found a wealth of baby *Dorudon* teeth. Most of the remaining whales were the serpentine *Basilosaurus*, but only adults. Perhaps Zeuglodon Valley was a calving ground for *Dorudon;* and perhaps *Basilosaurus* terrorized the young, just as killer whales today hound humpback whale calves, the survivors slipping away with flukes marked by moon-shaped bites.

When Gingerich returned to Zeuglodon Valley in 1989, he had a shopping list: he wanted to bring back flippers and, if he was lucky, hips. A *Basilosaurus* pelvis consists of two eight-inch slabs of bone with holes poked through them, joined together at one end to form a V. Most previous hips had been found split and scattered far from their skeletons, but with the wealth of *Basilosaurus* fossils in the valley, Gingerich hoped that one had hips in their proper place. For most of the season the dig was a failure. They found bones, but ones they could not name. While working on a little *Dorudon*, Gingerich found some fragments of unidentifiable bones, including one in the shape and size of a robin's egg. He thought that it looked like a kneecap. "But at the time it wasn't possible to think that," he says. After all, these were big, open-water whales. If any whale should have a knee (and the leg that it hinged), it should be something that still spent time walking on land, like *Pakicetus*. By the time whales lived in Zeuglodon Valley, a leg would have been an annoyance.

His search for hips was barely more than random. "The problem is, on a fifty-foot skeleton, where do you look?" Toward the end of the season Gingerich sat on the valley floor, working his way down a *Basilosaurus* spine that the wind had blasted flat and white. Near the forty-eighth vertebra, thirteen yards down the whale's back, he found a bone in the ground. "I was just hoping we'd find a pelvis, and then we found this," Gingerich says, holding up a baton of bone. "It looks a little like a rib, and yet when I looked at it, it doesn't look exactly like a rib, and it was way down along the vertebral column. I wondered if that was a femur. I wasn't looking for one—I thought they'd be as big as a ballpoint pen and about as interesting. And then peeking out near it there was enough bone to see that there was a knee, and there were articulations for a tibia and fibula. I got excited because these things came from a skeleton that was laid out, and so I knew where to look for more."

They had only a week left but returned to their many skeletons, working back to the first vertebra that fell under thirty centimeters in length, and fanning out for bones. The next day Gingerich had more luck. "I hit this bone with a shovel, and it went up in the air, scattering the sand," he says, twisting a little fossil in his hand and then bringing it down in his open palm as he tells me the story: an ankle. While they packed up camp, putting fragile bones in plaster jackets and striking tents, they still looked. On the last day Holly Smith swept away the sand from three perfect toes.

Back in Michigan, Gingerich tried to fit the bones together as they would have been in life. He could tell they were not like the vestigial bones that have been found in living whales, because they had all the marks where strong muscles and tendons attached to them. But when Gingerich tried to put them in a

normal mammal configuration he could only grind bone on bone. The knee
was rebelling. On a human leg, as on those of most mammals, there is a smooth
trough at the bottom end of the femur where the kneecap fits and slides. But
Basilosaurus's femur had only two recesses, with no accommodating groove in
between. The leg could sit comfortably only in two positions: one in which the
legs were held tight against the body, and another in which the leg locked open.
There was no middle ground. "I imagine they would go *boing!*" Gingerich says,
smiling. And now he was left with a riddle: what was a whale the size of a school
bus doing with legs the size of a child's, and ones, moreover, that could go
boing?

We think of what we are, what we have become with our souls and words, but
we don't think much about what we gave up along the way. Unlike a Wyoming
primate, we cannot ride the top of a forest canopy across a valley filling with

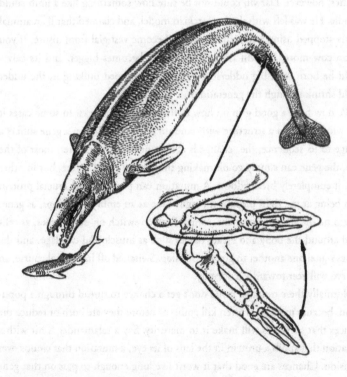

Basilosaurus, complete with legs (shown in their two positions)

night fog; unlike a fish, we cannot dive naked down an ocean gorge past waving violet sea fans. Species are not the only things that become extinct: bones and organs and senses disappear over time. We have lost gills, tails, and fur; our appendix, once a pouch full of plant-digesting bacteria, mocks us with the occasional case of appendicitis. Whales once had legs of some kind; *Basilosaurus* had delicately turned legs that rose and fell as it swam through the Tethys; and now, except for the rare stub, whales have no hind legs at all.

To Darwin, such shrinking vestiges were some of the best evidence for evolution. He was struck by how fish and other animals that lived in caves were so much like their relatives that lived in the light except for their pale, unpigmented skins and their sightless eyes; how birds and beetles that could not fly nevertheless had crude baby wings. If God had created all the species on earth as they are now, why should He leave these sloppy mistakes? Owen might say that these vestiges were actually the traces of an animal's transformation from the Archetype, but Darwin argued that an ostrich was not created with flightless wings—its wings had shriveled over generations. Without understanding genetics, however, Darwin could not be sure how something like a limb could dwindle. He was left with the Lamarckian model, and claimed that if an animal simply stopped using a structure it would become vestigial from disuse. If you milk a cow more, Darwin thought, its udder becomes bigger, and its calves would be born with big udders as well; if you stopped milking it, the udder would shrink through the generations.

We now have a good grip on how evolution's eraser works. In some cases it may winnow down a structure with random mutations. When a gene suffers a change to its sequence, the change is almost never for the better; most of the time, the gene can simply go on making the protein it did before, but in other cases it completely breaks down. A mutation can prevent some crucial protein from being at the right place at the right time as an embryo develops, as genes determine the shape of proteins, some of which switch on other genes, as cells crawl around the body and create tissues such as muscle and cartilage, and the tissues signal one another to grow or subside. Switched off its normal course, an embryo will veer toward a deformity.

Normally these mutated genes don't get a chance to spread through a population, because mutations often kill embryos before they are born or reduce the chances that survivors will make it to maturity. Say a salamander is hit with a mutation that spoils a protein in the lens of its eye, a mutation that clouds over its vision. Chances are good that it won't live long enough to pass on that gene if, like its parents, it depends on sight for survival. But if sight for some reason no longer has such a premium—if a population of salamanders finds itself catching insects in a cave, for example—the mutant salamander won't suffer and may have babies of its own. The mutation may randomly spread through the population as the generations pass by. Other vision-damaging mutations that strike aren't weeded out either, and the salamander eye degrades until it barely exists.

Yet this winnowing can't explain all of the weird things that happen when animals are born. Take the gristly leg that was found on the Canadian whale in 1909. How could a single whale reach back over thousands of generations of blurring, amputating mutations and reconstruct a crude limb? Such throwbacks are known as atavisms, and although they were once cast aside by biologists as meaningless freaks, it's now clear that they are actually natural experiments in evolution. Humans are sometimes born with stubby tails or with thick mats of hair covering much of their body, harking back to the Wyoming jungles. Horses have only three digits; the outer toes are slivers clinging to the giant middle toe that ends in its hoof. Yet horses are sometimes born with five fully formed digits. Greeks and Romans considered these rare creatures magical, and both Alexander the Great and Julius Caesar rode them. In each of these cases, an ancient program for an embryo's development turned out not to have been destroyed but somehow stored away, waiting to be turned back on again.

Over the past forty years biologists have even tried to trigger atavisms in the lab. Before 70 million years ago birds had teeth, which are the product of a union of two kinds of cells. The jaw begins as a piece of gum-coated cartilage. Meanwhile along the length of the still-forming spine a horde of cells called neural crest cells come into existence and begin to wander around the body, contributing to the formation of everything from nerve cells to voiceboxes to hearts. When neural crest cells reach the jaw, they issue commands to some of the gum cells, making them secrete enamel or produce other tissues of the teeth. Birds, like other vertebrates, have neural crest cells, but they can no longer grow teeth in their jaws. In 1980 scientists took some of this gum tissue from a chick embryo, combined it with neural crest cells from a mouse, and then grafted the tissue onto the backs of other mice. The graft sprouted into teeth, complete with crowns and roots. While the neural crest cells in birds can no longer issue the necessary tooth-forming commands to gum cells, the biologists concluded, the gum cells still retain the capacity to become teeth.

It is easy, though, to see atavisms where they may not actually exist. In a recent experiment, geneticists plucked a gene out of mice and observed that the mutant mothers would completely ignore their babies, going about their business as their infants died around them. The loving care of a parent is a relatively new behavior in the lineage that led to mice and other mammals: early tetrapods and synapsids probably guarded their eggs or hid them until hatching, but once born, the young were on their own. Does this mean that the negligent mother mouse has reverted back to a 250-million-year-old behavior? Not really. The gene in question appears to create proteins that serve as a single link in a long chain from the mouse's eyes and nose to the parts of the brain that drive nurturing behavior. When this link was removed, the chain was broken, but the other links still remained in the mouse's brain. Absence is not a good sign of the past—it can simply be absence.

And even when an archaic structure seems to reappear, it too may be an illusion. Bird teeth are a bit less wondrous now than they were when first reported. Some researchers have reported little luck when they tried to reenact the grafting, and skeptics now wonder if a few mouse epithelial cells slipped into the mix where teeth formed in the original experiments. In the early 1990s, geneticists captured the public's imagination when they tampered with the Hox genes in mice, and claimed that they had turned some of the mice's ear bones back to their old incarnation as part of the synapsid jaw. But anatomists have called their results into question as well. Cartilage, as we've seen in the case of bird shins, doesn't form only because a gene says it should. When embryonic tissues interact with each other—such as when a developing tendon is put under pressure—cartilage can spontaneously form. On the other hand, Hox genes may simply encourage more cells to congregate near the jaw than in a normal mouse, and the cells may form a new clump. That makes it hard to say that a new bit of cartilage that happens to show up in the neighborhood of an old bone is really an atavism.

Nevertheless, many atavisms are genuine. They are important both as markers of evolution and even sometimes as its agents. Consider a gene *A* that switches on another gene *B*. If a mutation knocks out *A*, *B* may live on in silence for a few million years in good working order. In that time a fresh mutation may bring *A* back into service or another gene may evolve to take its place. In either case, *B* suddenly can make its protein again, and if the protein normally triggers a cascade of growth in an embryo, its resurrection can be dramatic.

Some species of salamanders, for example, go through life as permanent children. They emerge from eggs underwater as larvae but never metamorphose into an adult shape, simply retaining their larval gills and living underwater until death. In some species of the axolotl in Mexico, this life of arrested development is the result of a single silenced gene, which normally produces a hormone called thyroxine. Inject one of these axolotls with thyroxine and you signal its cells to switch on a number of genes that ultimately transform it into an adult salamander. In other axolotl species the salamanders can take one path or the other, depending on some external cue like temperature. Axolotls invaded a region of Mexican highlands only 10 million years ago and have evolved into dozens of species. A cladogram shows that in that time arrested species have given rise to maturing ones, maturing ones have given rise to arrested ones, and both have given rise to species that can grow either way. In other words, the thyroxine gene has been turned on and off repeatedly in 10 million years. Not only do these salamanders experience atavisms, but they found new species as they do.

Atavisms from such silenced genes can turn up only within a few million years—if any more time passes they are ruined for good. A 40-million-year echo such as the stub of a whale leg is of a far older pedigree, and reveals a different way in which things disappear. The quadrupedal ancestors of whales built their limbs with the help of some of their Hox genes, but these genes are also involved in many other tasks—patterning the rear third of their backbone, for example, as well as their genitals. To get rid of their hind legs, you can't simply silence these genes because in the process you'd wreck their front legs (which became flippers), not to mention the back third or so of their body and their reproductive system. The development of a whale embryo shows how evolution chose a gentler course. Like other mammals, whales have a full complement of Hox genes that shape their spine as well as their four limb buds, the front pair of which continue growing into fins. The back pair get as far as forming bits of cartilage before the cell-killing program in the genes—which carves out our own fingers—gets an early start on them and kills the buds back to nothing. If a whale is born with a mutation that somehow weakens or delays the effect of the leg-killing genes, crude versions of limbs may form in much the same way they did 40 million years ago.

At the levels of genes and cells, biologists have a decent grasp of how evolution gets rid of structures—certainly much better than their grasp of how things come into existence for the first time. But it's not a simple trip from this biochemistry to the way those structures actually vanish during the history of life, with natural selection and other forces coming into play. As animals move into darkness—whether they are salamanders slinking into caves, bats flying into the night, or fish descending into an abyss—their eyes often actually swell rather than shrink as they try to sponge up the dying light. One particularly desperate example is a shrimp that lives at the bottom of the ocean, where vents form at the spreading ridges of tectonic plates and spew boiling mineral-loaded water. Somehow these vents produce a glow—perhaps by the energetic popping of boiling bubbles or the cracking of rock—that is invisible to a human eye. Yet light meters can register the dim fire, and so, apparently, can some shrimp that seethe along the flanks of the vents. They have turned their eyes into huge slabs of photoreceptors lodged in their back, good for nothing but detecting light. That, apparently, is all they need eyes for: by judging how bright the glow is, they can scuttle along the vent, perhaps to feed on the choice mats of bacteria, or simply to avoid being boiled. Eventually, though, animals encounter darkness so hopeless that they hit what biologists call the quit point. Beyond it, an investment in eyes brings so little dividend that they quickly dwindle to pinpricks, covered over by scars of skin.

Likewise, you might think that when a tetrapod loses its limbs, it's a straightforward process: it finds that life would be easier without legs, its legs dwindle to nothing, it reverts to its old fish's side-to-side muscle contractions, and it slithers away into its new existence. Yet limbs also disappear in counterintuitive ways. Snakes are only one of dozens of lineages of lizards that have lost part or all of their legs and now swim across the ground, through sand, or over forest litter. Although herpetologists still haven't figured out how many lizard species are related to one another, they can offer some of the best insights into how vertebrates lose their legs. If you line them up in series from full legs to none at all, some common patterns emerge.

While studying Australian skinks, for example, Carl Gans of the University of Texas found that they begin the transformation not by a simple leg amputation but by first elongating their bodies, apparently in order to make their way through crevices more easily. This is no small change in itself, since the lizards have to stretch out their organs, uncoil their intestines, and rearrange parts that were once side by side into a line. Once elongated into a serpentine body, though, they still aren't ready to give up their legs and start slithering. They pass through a stage where they move like walking concertinas: with their front legs planted, they take a few steps forward with their hind legs in order to fold up the trunk, and then the front legs proceed and straighten out the body. Walking this way through tunnels or burrows, a lizard ends up pressing the curves of its body against the walls. If it stiffens the muscles along its ribs at the point of contact, it can push against the earth and move forward. Now the animal can start evolving the reflexes that will let it move like a snake. It's not fish locomotion resurrected, though, but something far more complex: as a lizard bends its trunk it has to continually test each point at which it presses against the ground or a burrow wall, judging which line of ribs it needs to tighten to propel itself forward. Now, at last, lizards can give up their legs. What may look in hindsight like simple atrophy is actually a full-body reorganization with the atrophy coming as an afterthought.

Each kind of tetrapod probably becomes limbless in its own way. Caecilians are amphibians that look like blind snakes without scales, but according to work done by a student of Gans, James O'Reilly of Northern Arizona University, they appear to have evolved to such a similar shape along a completely different path. The difference stems from the fact that amphibians never turned their ribs into the muscle-bound cages that amniotes used for breathing. The four-legged ancestors of caecilians thus couldn't use them to stiffen their bodies to get traction against the ground. Instead, they held their breath and squeezed their muscles inward against their body cavities, and the high pressure made their bodies stiff.

With a few minor changes to its anatomy, the pressurized body of a caecilian could alter its shape. The attachment of skin and muscles loosened from its vertebral column (an easy operation, since its ribs are so small). At the same time evolution loomed the fibers in its skin into a cross-woven helix—the same geometry that dolphins may use to swim with a spring. At the angle that the fibers form in a caecilian's body, however, they have a completely different property: when they are placed under pressure they stretch, elongating the amphibian's body. And so when a caecilian pressurizes its body, its sides stiffen and its head is driven forward. It lets the pressure drop and pull its vertebral column forward inside its body, and then pressurizes again. This earthwormlike travel lets a caecilian generate twice as much force as a snake the same size. Perhaps most remarkable, the caecilian—which looks as if it is all tail—has lost its tail completely, since it is useless for this kind of movement. Instead, a caecilian is all gut.

Another common error when it comes to vestiges is to assume that because they are no longer their former selves, they've become useless. Blind mole rats live their entire lives in underground tunnels, feeding on roots and bugs. Fur and skin cover their eyes, which completely lack pupils or eye muscles and have only 823 retinal nerves (compared to 100,000 in an aboveground relative such as a hamster). Shine a light in a mole rat's eye and no sparks crackle in the visual cortex, the region of the brain where any other rodent would be putting together a mental image of the beam. But the blind mole rat needs its eyes. Its few retinal nerve cells may not connect to the visual cortex, but they do shoot dense branches into the hypothalamus, where mammals—including apparently mole rats—turn cycles of light and dark into a clock and a calendar with which to control their hormones, their waking and sleeping, and their mating cycles.

Given the same incentives for losing a structure—darkness, for example, or burrows—not all animals can give it up. A creature can only lose its eyes if it already has some type of sense organ ready to take their place. As a fish goes blind it can resort to its lateral lines, evolving them to be sensitive to the electric fields of other animals a hundred feet away. Deep-sea squid—animals from across the great invertebrate-vertebrate divide—never evolved anything remotely like lateral lines, and so they have no alternative to their eyes. Their eyes can thus only bulge farther and farther out of their head, at dark depths that a fish would consider well past the quit point.

And far from being crippled by evolution's eraser, an animal like a whale or a blind mole rat can be made fitter. It may be that in many cases, natural selection, rather than the random effects of undisciplined mutations, actively erodes part of an animal's body. Eyes are wasteful pieces of equipment: retinal tissue, ounce for ounce, swallows up a hundred times more energy than the average hunk of body, and vision demands vast spreads of real estate in the brain to process information. A blind mole rat, which uses its eyes only as sundials, needs less food and oxygen than an aboveground mouse, and it can convert the neurons in the visual cortex for more useful business, like sensing the roots its whiskers touch, smelling the urine of the other mole rats in its burrow, and listening to their coded foot-thumps. Gain and loss are the same to evolution: nothing more than change.

"You know," says Gingerich, holding a little thighbone of *Basilosaurus,* "these whales are fifty feet long. You could easily write off these tiny little legs as being vestigial—just waiting to disappear. But then you're struck by these surfaces for muscle attachment and well-formed joint surfaces." His thumbs move over the bone. "That doesn't sound vestigial. Then my thought was, 'Well, what does the pelvis do?'"

For over a century the United States Natural History Museum in Washington, D.C., has displayed an impressive *Basilosaurus* from Alabama with almost every bone accounted for, including its hips. Most mammal hips are like two halves of a ring joined together. They are fused at the spine (in the region called the sacrum), and it is through the sacrum that the legs' forces travel in order to move the body forward. The hips are also fused at the bottom of the ring, at a point called the pubic symphysis, where they anchor the muscles of the penis in males and the uterus in females. *Basilosaurus* hips are two flat straight pieces of bone, meaning that the ring had clearly been broken and had lost one of the two points of fusion. The museum curators decided that they must have still been connected to the spine at the sacrum and had pulled apart at the pubic symphysis.

These hips from Alabama were badly mauled by time and geology, though, while the bones that Gingerich found in Egypt were beautifully preserved. He could see the fine webbing that is the mark of two bones united, and from it he could tell that the fusion was not up along the spine, but at the bottom of the ring at the pubic symphysis. The hips floated deep in the whale's body, where they contacted the legs that Gingerich had discovered in Egypt. Obviously the legs would be of no use for locomotion, tiny as they were and separated from the spine. Yet they still had broad muscular attachments. Like the blind mole rat eyes, they might be used for another purpose. The pubic symphysis suggested to Gingerich what they might be good for: sex.

Not all snakes are totally legless. A few species of boas have little stubs that they use to guide their bodies into the right alignment for mating. *Basilosaurus* came about as close to a snake as a whale ever did, and Gingerich thought that kind of guidance might make sex an easier proposition. If that were the case, it would make sense for the legs to go *boing:* with the knees bent, the legs would tuck up against the body and the thighs would be scrunched inside the body itself, giving *Basilosaurus* a smoother profile as it swam through the Tethys. Locking into their other position, though, the legs would fit nicely along the flanks of another whale. "I think they are vestiges from the point of view of locomotion but not reproduction," says Gingerich.

A help for snaky lovers was how Gingerich first offered up *Basilosaurus*'s legs to the world in 1990. "But I don't think the snaky part is critical any more," he says. At the beginning of the 1989 season he had found the robin's-egg bone next to *Dorudon*, the smaller whale. At the time he had ignored the possibility that it might be a knee, but once he found a full leg of *Basilosaurus*, his mind opened to the idea. Other fragments his team found around *Dorudon* turned out to be parts of its femur and ankle. In 1994 Mark Uhen, then a student of

Gingerich's, found pelvic bones of *Dorudon* hiding in a drawer in the Charleston Museum in South Carolina. "It turns out that the pieces of the *Dorudon* hind limb are the same," says Gingerich, "but on a smaller scale for a smaller whale. Evidently it isn't the long serpentlike body form that is responsible for keeping a copulatory guide. They're just present at this stage of whale evolution and subsequently lost."

Before his work in Egypt, to see the heritage of whales on land Gingerich could look at only one fossil: *Pakicetus*. With his discovery of swimming whales with toes, he had come close to the end of the full transition of whales to the ocean. "All of a sudden a light went on in my brain: here we are 10 million years—*10 million years*—after *Pakicetus*, and these guys still have feet? I thought, 'We can study this problem.' Here's a transition spread out over more than 10 million years, plus it's in a setting with lots of fossils in shallow marine deposits." Between *Pakicetus* and *Basilosaurus* paleontologists might be able to find many whales graded from walking to swimming. And to find the walking whales, Gingerich had a feeling that someone would have to go back to Pakistan.

Chapter Eight

WALKING TO SWIMMING

Only once did Frank Fish have to live up to his name. About seventy miles northwest of Ann Arbor, where Phil Gingerich brought home his whales, is reedy Rose Lake. While Gingerich picked over the ears of *Pakicetus*, Fish was traveling to the lake to trap muskrats, and on one visit the burly, coal-eyed scientist almost, as he puts it, "became integrated with the ecosystem." He pulled up chest-high waders and sloshed slowly into the cattails, carrying with him a borrowed trap that he had sworn to return to its owner in pristine condition. That day he planned to visit a new muskrat lodge, but as he approached it the water rose quickly, slipping over the rim of his waders. Just as Fish decided to turn back, he discovered that he was walking on a four-foot false bottom. It gave way and he disappeared underwater. While the waders ballooned and pulled him down, he had a vision of himself found days later, cemented to the lake bottom with his trap still held just above the water, looking like some annual trophy for field work. The next thing he remembers he was on shore.

Most of the time Fish is a terrestrial mammal like the rest of us. He can swim a decent crawl, but he would need twenty times more energy than a fish of equal size to cross Rose Lake. These sorts of facts—the fundamentals of how mammals manage in water—go far toward defining Fish as a scientist. "If you ask me what I am, I have a hard time keeping it simple," he explained to me as we talked in his basement lab at West Chester University in the Brandywine Valley of Pennsylvania. He has a hard time keeping the lab simple as well: inflated dolphins and killer whales are suspended from pipes, a small crocodile gapes motionless in its tank, soot-colored horns hang on his wall, old televisions sit next to a long freezer where he sometimes keeps pieces of humpback flukes. "I say I'm a zoologist, but if people want to go beyond that, I call myself an ecological-physiologist-slash-functional-morphologist. That's just too long, but that's what I do." Labels aside, thanks in large part to Fish we now have precise measurements of the forces, energy, and movements required for mammals to swim. And although Fish doesn't dig fossils, he can also help show how macroevolution works by using living animals to explain how a trotting mesonychid could have become a swimming whale.

When Fish was five he decided he wanted to be a paleontologist, but by the time he turned seven he had realized that all the dinosaurs were dead and there'd be no fun in playing with them. At Michigan State University he stepped into the ecological physiologist half of his future self—the kind of scientist who, among other things, measures the flux of gases and heat in and out of an animal during its natural activities. He dabbled a little with fishes, but the field was crowded with experts and as a result many of the big questions had already been settled. But few had studied swimming mammals carefully, despite the puzzle of how they used a body sculpted by hundreds of millions of years of evolution on land to move through water.

For his master's thesis Fish wanted to study how a warm-blooded mammal adapted its metabolism to water, which sucks away heat like a thermal vampire. He scanned Michigan for a subject. Beavers—enormous beasts that could slap half the water out of an aquarium with their tails, clamber over its wall, and wander out of the flooded office—were too much grief. Water shrews generally died by the time they arrived at the campus. The best Fish could hope for was muskrat, not that he has much love for them. "Vicious little monsters" is the term he uses, respectfully. After he caught them at Rose Lake and brought them back to the campus, he used nets and brooms to usher them from trap to tank. Fish measured their body temperature in water and air, in warmth and cold. His experiments showed that muskrats use the same basic strategies as dolphins, managing their heat by steering the circulation of their blood. A muskrat's slender, hairless tail and legs have a such a high ratio of skin to flesh that they can radiate heat much faster than the rest of its body. At eighty degrees, a dry muskrat's tail grows hot and dumps enough warmth to keep its core from baking. At cooler temperatures it needs to hold on to heat rather than release it, and so it shuts its tail circulation down. When Fish brought the tank down to sixty-four degrees, the muskrats kept their bodies warm but let their tails match the ambient temperature. In water, on the other hand, which can draw off heat far easier than air, the muskrats always kept their tails choked off, even when Fish turned the water temperature up to eighty degrees.

Physiologists sometimes get into the habit of thinking of their animals as nothing more than boxes, with energy and matter flowing into and out of them. But as Fish worked with his vicious little monsters, he thought about how they not only adapted a warm-blooded metabolism to water but managed to use mammalian limbs to move through it. As the functional morphologist half of Fish emerged, he discovered that the hydrodynamics of muskrats was another well of confusion. "In the literature, muskrats swam with their tail alone, or using their hind feet alternately, or hind feet together, or front feet without the hind feet, or all four legs—and everyone who's saying these things is looking at these animals as they're swimming along in the river. They can't even see the feet."

To make a careful study of their swimming, Fish put his muskrats in sealed tanks of flowing water and filmed them traveling against the current. Projecting the film frame by frame on a screen, he drew stick legs, and from these he could calculate the forces the muskrat generated with each stroke; by measuring how much oxygen the muskrats consumed, he could see how efficiently they turned energy into movement. Muskrats turned out to swim across the surface of water by paddling only their back feet, holding their front paws motionless close to their bodies. As they pushed each rear paw back, they opened their hairy, webbed toes wide in order to ram the greatest surface area against the water. Once a leg had kicked back as far as it could reach, the muskrat closed the paw and drew it forward again for another push. Swimming this way, muskrats burned about as much energy, pound for pound and mile for mile, as a human—an aquatic bonfire.

Fish visited zoos. He filmed otters, dolphins, seals, polar bears, any mammal with its own pool. Almost all mammals can swim—the lore is that only apes and giraffes can't—but few have been studied carefully. For a few years Fish could be satisfied to draw his stick-figure legs for species after species, even if he wasn't sure where the work would take him. "For a while I was adrift," he says. "I didn't know what to do with the muskrat stuff other than to repeat it, and there was no guiding model for doing it. It was stamp collecting."

Once he had amassed enough movies, Fish almost unconsciously began to categorize his swimmers, laying them out along a spectrum. At one end were the majority of mammals, citizens of terra firma that could get across a creek if need be, slowly and inefficiently. Toward the middle were species that could dive for clams or roots, and at the other side the mammals that came ashore only to mate or not at all, that swam as easily as other mammals walk. The mammals in each category had much in common in the ways they swam, in their anatomy, in their efficiency, and in the importance that water held for them—despite the fact that they came from distant branches of the mammal tree. It occurred to Fish that he could use his movies not just to describe the swimming of living mammals but to investigate the macroevolution that produced it in the first place. "I started thinking about the basic question: how do you go from a terrestrial quadruped to a dolphin? Why do you do that? And why do you get other morphs, like seals and sea lions?"

Fossils probably couldn't help him much, Fish decided, in part because of their scarcity and in part because of the way paleontologists toyed with them. When a report on the back of *Pakicetus*'s skull was published in the prestigious journal *Science* in 1983, a full-blown portrait appeared on the cover: a beak-nosed, bearish creature with stubby paws, plunging deep and far after herring. Fish was not pleased. "The paleo side I always felt was—to be kind, I'll say less rigorous. They're much better now, but at the time we had this incredible amount of storytelling. You have bones, you reassemble them, and then you can argue anything you want. When I saw the *Pakicetus* I was put off quite a bit. It might be true, but there was absolutely no evidence for this. This was beyond even just a reconstruction where you have a full skeleton and you can argue whether *Tyrannosaurus rex* had its tail on the ground or didn't. At least we knew it had a tail."

Fish couldn't study the evolution of cetacean swimming in the way it's possible to study the evolution of tetrapod breathing. The transition from water to land is embodied in the node-to-node path of the cladogram on page 103. Ray-finned fishes, lungfish, salamanders, and lizards are living members of the lineages that branch off these nodes, and although they have since evolved in different directions, their breathing doesn't seem to have changed much. But when Fish looked to a cladogram of whales and their relatives, the closest thing on land he could hope to study would be an artiodactyl such as a moose—an animal that had diverged far from mesonychids in the past 60 million years. "You can throw those animals in water, but you're not going to see the same thing. They're all adapted to speed or climbing rocks or something like that," says Fish.

At least Fish knew that an ancestral mesonychid did not leap off a cliff and hit the water swimming like a dolphin. By sorting through his categories of swimmers he could generate a hypothesis about the stages through which a terrestrial mammal could ease into an aquatic form, with behavior and anatomy playing a gradual game of leapfrog. When paleontologists did finally find some transitional fossils of whales, he could put his scenario to a test. But as Fish settled on this plan he recognized how many categories he lacked, how many animals he needed to watch. He needed to put opossums, both North and South American, into his tanks, as well as platypuses and beavers, rice rats and Australian water rats. He crouched on catwalks to film killer whales tearing by, he stood by glass-walled tanks to record different species of dolphins rushing past, and he filmed otters of all stripes: sea otters, river otters, and the giant otters of the Amazon.

By the early 1990s he had enough footage to put together a theory of how whales came to swim. Step one would be a dog paddle. Dogs and other terrestrial mammals (including, presumably, mesonychids) all use an identical stroke in water: the left foreleg pushes back as the right hind leg comes forward, and then the right foreleg pushes back as the left hind leg comes forward. A dog paddle is nothing more than a trot thrown in water, and requires nothing more from the brain than the same motor patterns needed for walking on land. For an animal that wants to swim only in order to get across a creek, it will do.

Compared to other swimming styles, however, a dog paddle is feeble. To slip through the water you want to move like a torpedo, presenting as small a cross section of your body as possible. But a paddling mesonychid with a chest full of air would tilt up as it swam. Another inevitable woe is the way dog-paddling legs interfere with one another. "We had possums swimming," says Fish, "and as one leg would go back and one would go forward, you'd actually see them collide and the animal would stutter like a horse when it trips. And hydrodynamically, you're pushing water and applying momentum to it, so that you're changing it from the way another foot would experience it if it were going through clean water."

It takes only a few adaptations to spruce up this swimming stroke. As mesonychids spent more time in and around water, they could have experienced a shift that Fish found when he compared the North American and South American opossums. The North American opossum, which is more likely to be in a tree or staring dumbly at an oncoming car than in a river, is a standard dog paddler. Its more aquatic southern relative (known also as the yapok) looks almost identical, but it holds its front legs out in front of its body while it kicks with its hind legs so that it no longer has to cope with interference. The yapok's hind feet are webbed to add thrust, its eyes ride higher up on its head, and its fur traps air bubbles to give its entire body buoyancy. While the North American opossum struggles and bobs as it swim, the yapok keeps a horizontal trim and paddles easily.

That's about all evolution can do with a dog paddle, though. At its heart, it's a two-phase cycle: each leg goes through a power phase as it pushes back against water and then a recovery phase when it pulls forward again. Every recovery slows the animal down, even if it can reduce the drag by closing its paw. Making matters worse, a dog-paddling mammal swimming on the surface has to fight against the turbulence of small waves. It creates it own bow waves like a ship, which draw energy away, and as a ship or a swimmer moves faster, it reaches what's known as hull speed, at which point it becomes trapped in the trough of the wave and can go no faster.

Whales could not have moved beyond this inefficient yapok stage of evolution, according to Fish, without crossing over to a new swimming style. They had to become, for a time, like otters. The body of an otter is substantially different from that of a land mammal: it has a long, muscular tail, a cigar-shaped trunk, and stumpy legs; its front feet are small and its back ones are oversized. While traveling across the surface of water, it is still able to paddle, but in order to dive for its food, it needs a new style. If it tried to descend by swimming like a yapok, the buoyancy of its fur would overpower the modest thrust of its hind legs.

Rather than alternating its hind limbs, an otter pushes them back together during a dive. It is still relying on the basic principle of pushing water that dog paddlers use, but it can now also get thrust from its tail. Its flexible, long back bends as it kicks, so that when its hind legs have finished pushing and start to pull forward—normally the point at which the animal would lose its forward momentum—its undulation continues into its stout tail, continuing to push against the water. By the time the wave has traveled to the tip of the otter's tail, its hind feet are curled up for another kick. The combined effects of this powerful stroke and the liberation from the tyranny of surface waves are immense: a sea otter can swim 75 percent faster underwater and cut its cost of transport almost in half. And yet to swim in this way, an otter has no need for a new wiring for its brain: instead of using a terrestrial trot, it uses the same motor patterns that generate a gallop.

An otter is still a long way from a whale. "You're bobbing on the surface like a cork. It's going to take a lot of effort to drive yourself down," says Fish. Whales would have needed to lose their air-trapping fur at this point and evolve blubber. "By changing from air to something like blubber, you get good neutral buoyancy. You only find sea otters in shallow water. They're not in the open sea. Blubber you don't have to maintain—otters spend a lot of time grooming their fur. And blubber's an energy source you can store up and live off." A blubbery proto-whale could continue to swim like an otter as its tail grew larger and more muscular, its back even more flexible as it was loosened from its hips. It would rely less on its legs because it could generate enough thrust from flexing its back without the initial push from the hind feet. The last step to whale swimming would be the addition of flukes to their tails. Now, with these airplane wings on their tails, they would be able to generate lift, which is the most efficient way to move through any fluid, be it air or water. And by adjusting the angle of the fluke, they'd be able to propel themselves as far with every upstroke as with every downstroke.

What makes Fish's model particularly powerful is that it explains not only how a whale could have evolved but how other marine mammals ended up with

An otter kicks with its hind legs and generates extra thrust by bending its back and tail.

their own anatomies. Seals could also have descended from dog-paddling animals that stopped using their front legs and swam like yapoks so as to cut down on interference. But then they made a different choice: rather than pushing back with their hind feet, they angled them to the sides like oars—a stage embodied, as Fish discovered on a trip to Australia, by the platypus. They began to row their feet from side to side, and gradually the entire pelvis evolved to oscillate as well.

The separate fates of seals and whales were not arbitrary, however: they actually bring Fish back to his first work as a zoologist with tails. A tail can quickly unload the extra heat in a mammal, which makes it an asset in the tropics, but in colder climates any extremity is a liability. "If you take a look at possums up in the north and the south," Fish says, "you'd think they're two different animals. Possums in the north are always ratty, with ears that look like they've been bitten off, tails are half missing: frostbite. Possums in the south, they don't have any trouble like that because they don't experience frostbite, so they have big floppy ears and a long tail and all their toes—and they look somewhat beautiful." Seals, the fossil record shows, descended from a bearlike animal at least 25 million years ago on some chilly coast of the northern Pacific. This animal had no tail to speak of. The oldest whales, by contrast, come from Pakistan, which was close to the equator 50 million years ago, and a long mesonychid tail could have helped keep it cool. Thus their homes on land determined their futures at sea. William Flower had suggested over a century earlier that tails had decided who would inherit flukes, and with a deeper knowledge of physiology, Fish has ushered the idea into modern evolutionary theory. As Fish says, "You're stuck with your history."

In 1979 Phil Gingerich discovered *Pakicetus,* the oldest whale; ten years later he found the youngest whales with legs; and in 1992 someone opened a volume for the first time between these two bookends. It would be a clichéd plot twist to make that person a student of Gingerich's, but that's what happened, and this story has to make do with the facts, no matter how pat. When I visited Gingerich's former student, Hans Thewissen, at the Northeastern Ohio Universities College of Medicine, he was spending an afternoon standing on a table, pointing a 35-millimeter camera at a pillow of brass-colored rock between his feet. An assistant held a clip light, and Thewissen asked me to hold another to rub out the last shadows. We looked like a fashion shoot for a rock hound magazine.

The rock itself was a recent discovery from Pakistan and crammed with bones. I would have thought that Thewissen would be tearing it apart with his bare hands: sitting squarely at the top of the rock was a skull of *Pakicetus*, with the front half intact, providing for the first time a view of what its whole head really looked like. The tiny orbits of its eyes were exposed, bunched together high on its snout and far forward. Also jumbled in the rock were the fossils of turtle shells and of the little elephant uncles called anthracobunids. Pieces of shoulders and forelegs were mixed into the batter as well. They might belong to *Pakicetus*, which would mean that Thewissen had found the first bones beyond the skull of this primordial whale. Or not.

But Thewissen (his name is pronounced TEH-vissen) was taking his time. He jumped off the table and pulled the rock's mate toward it. As he worked, his expression was so intense he looked almost melancholy. His hair was cut flat over his forehead, his lips pursing his pale cheeks. Watching him, I thought of the solemn faces in Holbein drawings. He wouldn't let his assistant start extracting the bones from the rocks until he had photographed them from all angles. There is information in fossils, but there is other information in fossils that are still in rock. They get to their final resting place when animals die together in sandstorms, when bears build heaps of them in their favorite cave, when their floating carcasses slip out of main currents of a river and spiral in an oxbow. Thewissen sledgehammered these rocks out of a Pakistan hillside, and he wanted to use the photographs he was taking to draw a map of bones, one that might tell him how these animals died. They might indicate whether the unclaimed shoulder and leg really did belong to a whale. That was his hope, at least, as he tried to turn over one of the rocks in order to remember how they had originally fit together in a hill on the other side of the world. "Sometimes I wish I had taken a course in masonry," he grumbled.

"I used to try to get Hans interested in whales, but I couldn't," says Gingerich. Like Gingerich, Thewissen has been in love with fossils for a long time but has only recently been seduced by the origin of cetaceans. "I've been interested in fossils all my life, my mother tells me," says Thewissen. On weekends his father took him to an outcrop thirty miles from their home in the Netherlands where he could dig out traces of 70-million-year-old marine life. "He was interested because I was interested; he was mainly trying to keep me out of trouble. And I was an only child, so I got to pick where we would go on vacation. It was always to fossil spots, great Devonian localities in Germany with trilobites and corals, and for several seasons to the north coast of Normandy for the

Jurassic-Cretaceous transition, and a lot of vacations in Spain where there are Carboniferous mollusks."

By the time Thewissen came to study with Gingerich in Michigan, he had come to understand that he was a paleobiologist rather than a paleontologist. "I was more interested in the animals, not so much what happens to them after death, or how to use them to date the rocks." It is a rare paleontology student who spends a month dissecting an embalmed dog in order to learn what every tendon, gland, and muscle looks like. While at the University of Utrecht he studied artiodactyls and came to admire their ankles. In us humans and most other mammals, the ankle swings the foot at one pivot. But a buffalo, a moose, and all other artiodactyls have elongated ankle bones that can swing at both ends, giving them a long, stable stride. To understand the evolution of this double pulley he cut open aardvarks that died at the zoo in Amsterdam to study their primitive ankles, and he studied fossils, working at the National Museum of Natural History in Paris and helping out on a dig in Pakistan.

At Michigan Thewissen studied the condylarths, those primitive plant-eaters that would give rise to today's hoofed mammals, and traveled with Gingerich in the summers to find fossils in Wyoming. At the time Gingerich was spending his winters in Egypt and bringing home heaps of whale bones that he needed someone to study, but Thewissen didn't volunteer. He was more interested in mammals that still used their legs on land. After graduate school Thewissen taught anatomy to medical students at Duke University, where he spent time investigating the flight muscles of bats to find clues to their relationship to other mammals. He mulled over how he could push his career forward and decided he needed to find some fossils of his own. Pakistan was a possibility. Many of the questions that had brought Gingerich there were still unanswered. Geologists knew that the Indian subcontinent had plowed into Asia, closing the Tethys and creating the Himalayas, but they were surprisingly unsure as to when it first made landfall and how long it took to seal up completely. Once they had suggested 25 million years ago or so, then they pushed it back to around 40 million, and a few were looking curiously back to 65 million. The timing could have enormous significance.

David Krause, a paleontologist at the State University of New York at Stony Brook, looked over the mammal fossils that had been amassed since the 1970s and suggested that India had been the cradle of many of the living mammal orders. Before 70 million years ago, India had been sutured to the southern coast of Africa. Perhaps primitive mammals boarded the landmass before it embarked across the Indian Ocean, and during the next 15 million years, the mammals evolved on India in isolation from the rest of the world's animals. A lineage of lemurlike primates had evolved into a new form that would ultimately give rise to tarsiers, monkeys, apes, and ourselves. The condylarths produced some of the first hoofed mammals. When India made Asia its new home, these mammals raced off the gangplanks and traveled around the world in a few million years, turning up in fossil-rich areas like Wyoming. Yet as interesting as ideas like Krause's were, they relied entirely on the timing of the collision, which was still not pinned down. It seemed to Thewissen that a paleontologist could help. Before the collision, the Indian subcontinent would have been an island full of oddball life, and if he stood in Pakistan, at the northern edge of the subcontinent, he could watch the mammals of India pour out into Asia and the Asian creatures clomp their way in the other direction. And if he could date their fossils he might be dating the collision.

Thewissen got back in touch with the paleontologists who had hosted him in Pakistan as a student, a team led by researchers from Howard University and the Geological Survey of Pakistan. This was the team that had worked at Ganda Kas a few months after Gingerich in 1977, and in the late seventies they had found fossils of many mammals there, including a piece of whale jaw that Gingerich later renamed *Pakicetus*. The Howard-Pakistan team suggested that Thewissen go back to one particularly mammal-rich site called Locality 62, and Thewissen asked the National Geographic Society if they'd be interested in funding the project. "I thought my chances would go way up if I asked for very little money. So basically it boiled down to three tickets to Pakistan," he recalls. They gave him the money, and on the first trip in 1991 he took with him Andres Aslan, a sedimentologist who had worked with him in Wyoming. In Islamabad they joined Mohammed Arif, the assistant director of the Geological Survey of Pakistan, and headed for the Kuldana Formation, back to the region around Ganda Kas where Gingerich had strolled from his waiting taxi.

Arif, Aslan, and Thewissen would drive from the nearby town of Attock each morning in a battered blue Isuzu truck to a valley in the Kala Chitta Range. They walked a mile from the road to Locality 62, where Eocene siltstones, mudstones, limestones, and sandstones all stood exposed along ten-foot outcrops among the scrub. Thewissen needed Aslan's sedimentary wisdom badly here. They were walking backward through time, watching the Tethys Sea rising and falling so quickly that a few steps could take them into a completely new habitat. "The Kuldana is amazing in how quickly the environment changes," says Thewissen. "You've got freshwater rocks, then there's a marine incursion, a bay, another lake, then it's full ocean." And he needed to keep to the terrestrial rocks, for only in them could he hope to find mammals that were crossing the Indian frontier.

As the sedimentologists scanned the rock faces at Locality 62, Thewissen smashed them with a sledgehammer. After a few days he found bones, but from a whale ear—what Thewissen would later realize was the ear of *Pakicetus*. He knew that this was an important fossil, but not for his promise to tick off the mammals that walk into Tibet. "When you're looking at land mammal migrations, you're not looking for whales," he says. They worked for six days before they had to return to Islamabad, where Thewissen had promised to pick up a Dutch paleontologist flying in to study Miocene rocks. He was ready to head back out to the field when history stepped in. The American embassy told him to stay in the capital because the stand-off between the United States and Iraq was making the entire Muslim world tense. For a few days Thewissen bided his time in Islamabad. "Then we realized the negotiations were going nowhere so we just left. At least I got my six days. Plus we found those whale ears."

The ears kept him occupied for a few months back at Duke. Inside the grape-shaped shell of the ears he found the first known *Pakicetus* incus, the middle bone of the ear-bone chain. When whale embryos first form their ears, this chain looks basically like those of other mammals. But in the case of whale ears, Haeckel was right: ontogeny does recapitulate phylogeny. The incus twists about ninety degrees, pulling the surrounding strands of connective tissue around with it. The turn probably has something to do with adapting an airborne ear to underwater acoustics, but no one can yet say exactly what that something is. In any case, Thewissen found that in its incus, *Pakicetus* was once again a remarkable intermediate: compared to the ear of a living whale, its incus twisted only about half as far.

Thewissen's exile from Pakistan was much shorter than Gingerich's. By the next field season the Gulf War was over, and the State Department didn't interfere when he made his arrangements to return. His money was almost gone, though, and as he traveled alone to Islamabad he thought about how this was his last chance to find a trace of a migrating mammal. Arif joined him in Islamabad and they drove the Isuzu back to Ganda Kas. They headed back down the same valley to Locality 62. They didn't quite walk a straight path to the site; paleontologists have a habit of wandering unless they're weighed down with sledgehammers, picks, chisels, and shovels for a day of quarrying. As Arif strolled along one side of the valley, he spotted a patch of bone. He hacked the rock around it, and once it was free Thewissen recognized it as the rib of an ancient sirenian—a member of the group of mammals that includes manatees and dugongs, mammals that returned like whales to the sea not long after the dinosaurs died. Paleontologists suspect they arose from the elephantlike anthracobunids 50 million years ago, carrying their peaceful grazing from land to water where they fed on sea grasses.

Like the ears of *Pakicetus* it was a fascinating fossil—the first sirenian from the Eocene of Pakistan—but once again Thewissen was running into water when he wanted to find land. When he and Arif got back to Locality 62, they still couldn't find any good terrestrial mammals, and Thewissen decided it was time to strike out for fresh rock, to go prospecting. "You're just wandering," says Thewissen. "You see on an aerial photo an outcrop of the right age and the right look, and you prospect those areas, which means you spread out and don't see each other till lunch, and you walk slowly from bed to bed, not very directed, to see if there are any fossils there. That's how we were that morning, just seeing if there's anything in this place, and then we got together for lunch."

They ate leaning against a wall of wall of green and pinkish stone. It was also part of the layer in which Arif had found the sirenian rib—a few million years younger than the rocks where *Pakicetus* was found—and when Thewissen happened to look up at it, he noticed the rock's bumpy texture. Oyster beds that had once thrived just offshore in the Tethys Sea had sculpted them. Still sitting there eating, Thewissen found three fish teeth. It was as if the rocks were trying to tell him that he should give up trying to be India's timekeeper and consider what the sea could offer. He decided to listen.

Within a few days of digging in the ocean rocks he knew he had made a good choice, because he and Arif had found pieces of a sirenian's skull. A few days later Arif found a knee of something emerging from a low cliff of siltstone. "I recognized that it was a mammal knee because of the shape of the femur and the tibia," says Thewissen. "At that point I didn't know what it was."

When they found more bones of this mammal scattered near the knee on the surface of the cliff, they decided to quarry the rock to see how far back and down the fossil extended. "More and more bones came up. It's a vertical cliff, and we were walking on top and that's where we found the knee. Then once we identified which bed it was in we just went down, and the first place we hit there was a rib with a little foot bone by it. Then there's a femur there to the right and more ribs, and you start to run into the hand." The animal, something over six feet long, was buried head down in the rock, and as they exhumed it over the course of three days, Thewissen tried to guess what they had found. He thought perhaps it was an anthracobunid—a fair guess, given that nobody had ever found anything more than a jaw from an anthracobunid before. Maybe he had found a missing link between anthracobunids and sirenians. Still, it was a freakish body for any mammal to have. The forelegs were short but had great flat hands, while the hind legs ended in feet shaped like clown shoes. "I was thinking, it's a nice skeleton, but what am I going to do with it if I don't know what it is?"

They dug down deep into the cliff, and the last bones they found were from the skull. It was a long-snouted thing, broken up by the root of a shrub. They picked away the rock from one of the animal's teeth, the one thing about anthracobunids that Thewissen knew for sure. "Once I saw these teeth, I said, 'They don't look like sirenians.' " In fact, they had the mesonychidlike look of a primitive whale. He and Arif scraped rock from the ear, and saw that it drooped like a grape from behind the jaw, and then recognized the S engraved on it, the mark of cetaceans. It finally dawned on Thewissen: they'd found a whale that walked.

Thewissen did not return home triumphant, at least not immediately, because he was now broke. "I couldn't bring back the whole skeleton because I didn't have money to pay for excess weight," he says. At the Geological Survey in Islamabad he gathered the fossils he had found that winter. His other finds included the sirenian rib and skull, and a jaw of *Pakicetus*. This jaw would turn out to be important as well. Like the incus he had found in 1991, this jaw was beautifully intermediate. Land mammals—humans and mesonychids included—have a hole the size of a matchhead about midway along the inner side of each wing of their jaw. Through it flow blood vessels and nerves that keep the chin alive. In a whale, this hole is a wide slot that runs through much of the jaw; it houses the fat pad that funnels sound to their ears. In *Pakicetus*, Thewissen found, the hole was still small, while in the walking whale, about three million years younger, it became bigger. But Thewissen wouldn't scrape away the sediment on the jaw and see the hole for another year. In Islamabad, he had to lay almost all of his fossils in straw-bedded crates and hammer them shut. He could take home only what he could carry on to a plane, and he chose the skull of his new whale, filled with its taxonomic Talmud. He wedged wet toilet paper into its crevices, covered it with aluminum foil, and wrapped it in plaster bandages. It went into his backpack, and he went home.

Back in America Thewissen had to tell the same story again and again, each time with different shades of embarrassment: I've found a whale with legs, one of the most important fossils in the history of paleontology, but I don't have the legs with me at the moment. "He found it and then he called me," says Gingerich. "He called me to tell me how the season was going and for recommendations for a job. And I remember saying, 'Hans, how do you expect to get a job if you leave your best fossil in Pakistan?' Like me, he was interested in teeth and skulls, and when he told me he lugged this huge heavy skull on carry-on in an airplane, and left the box of limbs, I couldn't believe it. 'Don't leave your most spectacular stuff there! Get busy and work on it!' " Thewissen went to scientific meetings and tried to impress people with the skull and pledges that ten thousand miles away he had legs from the same whale. "I should have waited, but I was so excited—my God, I've got this amazing fossil, this missing link!—but I think most of the crowd was saying, 'Yeah right. We'll believe it when we see the legs and feet.' That's how I would have responded to somebody else."

It was Gingerich who bore the bones back to Thewissen. After his 1991 trip to Zeuglodon Valley, he was wondering how much more he could do in Egypt. He had found five different genera of whales, which his student Mark Uhen was sorting through. He had found legs and toes on a whale. He had even used the Global Positioning System of satellites to mark the location of every one of the 349 whale skeletons he had found in the valley so that later paleontologists would not have to rediscover the bones. If someone else wanted to continue the work, they were welcome to it. And with the end of the Afghan civil war, his thoughts turned back to Pakistan. He remembered that disastrous first field season, when he had brought an international team to the Rakhi Nala Valley to dig up approximately 43-million-year-old mammals, only to discover that they were standing on what had once been the sea. "Before we fled, we had things like this," he told me at his office as his hand glided up a bank of drawers, stopped, and pulled one open. Inside was a fused mass of cider-colored bones. "That's clearly the acetabulum of a pelvic bone. We've got a sacrum, and the backbone is fused where the pelvis attaches. We joked about walking whales, but it was just a joke then." They assumed instead it must have been some relative of an elephant that had drifted out to sea and rotted away into pieces. Now that he had found legs on *Basilosaurus*, he couldn't be so sure anymore.

As Gingerich prepared to return to the Rakhi Nala Valley in 1992, Thewissen told him stories of the whale legs he had left behind in crates. When Gingerich got back to Pakistan after his nine-year hiatus, the differences weren't many; the old British rifles that men carried in the countryside had all turned into Kalashnikovs. Once he had reached the Rakhi Nala Valley, he prospected for days at the northern end with Mohammed Arif and Xiaoyuan Zhou, a Michigan student. The rocks here were probably a few million years younger than the ones that Thewissen had been working on. One morning they stood on the crest and agreed on how they would break up and then gather again at a sharply pointed hill in the distance for lunch. Arif hadn't gotten far from Gingerich when he found bone, and the two of them spent the morning on it. It was a remingtonocetid, a narrow-faced early whale whose skull had been found in India. They met at the hill and in the afternoon split up again. This time it was Xiaoyuan who called Gingerich. "Along a little hill he found a few pieces of white bone. It clearly was there in the ground. I got excited. I thought, 'This is just like Egypt. It's just starting to come out—and it's the jaw of a whale.' I thought, 'This is going to be good.'

"I sat down for the next few days and started to excavate this thing. Almost immediately I hit the femur—it was almost curled around the front, so when you excavated the jaw you got to the femur." Gingerich had never seen anything like the thigh of this whale, only seven inches long. He remembered how Thewissen had told him about his own specimen, the size of a sea lion, and its eleven-inch thigh. This Rakhi Nala Valley whale—a little younger than Thewissen's—had far stubbier legs for its size, although of course they were strapping compared to the younger *Basilosaurus* legs he'd found in Egypt. The femur that Gingerich was now digging free had smooth patches at each end for healthy muscle attachments, and the groove for its knee was normal, lacking the grotesque two-position locking of *Basilosaurus*. And while the legs and hips of *Basilosaurus* floated alone in its body wall, this new whale had them close to its spine.

Gingerich found four other species of whales on the dig, but nothing could compare in abundance of bone with that curled specimen. In February 1992 he came back to Ann Arbor, bringing Thewissen's fossils as well as his own. For two years they each scraped the tenacious rock from their whales, and in 1994 they published their discoveries within a few months of each other. That year marked the first time that whales were no longer divided from other mammals by a wide rushing stream: two stepping stones had risen above the current.

Thewissen's creature, which he called *Ambulocetus* ("walking whale"), was the closer of the two to the origin of whales. Its four-hundred-pound body—an enormous crocodilelike head, a wide chest, and a long tail—sat on squat legs. It still had the tall projections rising from its neck vertebrae that mesonychids had used to hold up their heavy heads. The width of its chest pushed its hands

Ambulocetus

out to either side like seal flippers, and the giant feet on its crouched hind legs slapped awkwardly on the ground. *Ambulocetus* could shamble on land if it had to, but the shape of its spine told Thewissen where its gifts lay. It had lost the locking tabs that kept mesonychid spines rigid, and its general geometry was closer to an otter's than any other animal's. Although Thewissen did not find *Ambulocetus*'s hips, the spine strongly suggests that *Ambulocetus* could have

Rodhocetus

arched its back as it pushed out its giant hind legs and driven the force of its kick out to the end of its tail.

Gingerich's whale, which he named *Rodhocetus,* was probably somewhat more like a modern whale, although its head was still massive and was held up by muscles anchored to the same tall, mesonychidlike neck vertebrae. Although he doesn't know what the feet of *Rodhocetus* looked like, its legs are so short that he doubts that they could have generated any significant force in the water. Its spine, no longer fused to its hips, could now bend effortlessly all the way up to its rib cage, so that for the first time whales became essentially half tail. While *Rodhocetus* could still drag itself on the shores of the Tethys, it now depended completely on tail-drive to move through the water. Gingerich found only a few bones past its pelvis, preventing him from knowing whether it had grown flukes at its tip. But given that the shales where he found it were a fair distance off the coast, it must have swum from shore with a powerful stroke and must have only rarely come back to land.

Taken together, these two whale fossils were like an arrow pointing from land to sea. And perhaps most remarkable of all, they could have been designed from the specs that Frank Fish had been offering only a few years earlier, based on his studies of living mammals. Here were otter and post-otter at last made flesh, or at least bone.

Chapter Nine

A VOYAGE OUT

*A*mbulocetus alone cannot say how whales learned to swim, not even if *Rodhocetus* chimes in. Like any of macroevolution's works, the origin of whales needs a chorus of fossil voices to describe it. Fortunately, though, these two fossils have enough company. In the past few years a number of early archaeocetes have been discovered, and together they now tell a surprisingly coherent story, as strange as it may sound in the telling. Some parts of it are fuzzy, others are based on scrappy evidence, and still others suffer from acute contradiction. Its many authors are not ready to sign off on the entire story, but perhaps it can be history's first draft. At least now we have a draft.

Like most stories about macroevolution, this one begins in death. Around 65 million years ago, the way the world had worked for 150 million years came to an end. The dinosaurs on land, the pterosaurs in air, the marine reptiles in the sea all laid down their final bones, and for 10 million years life on earth was relatively quiet and small. Many old vertebrate survivors saw little change to their days: the sharks still roamed, the turtles crept on land and swam in the ocean, the crocodiles lay in wait. But nothing could yet reclaim the great sizes of the extinct creatures, their speed on land and at sea, their social graces. Toward the end of the Cretaceous period sea levels had been falling, rains slowing, and temperatures dropping, but in the period after the extinction—the Paleocene—earth warmed again. Palm trees swayed on Spitsbergen, not far from the North Pole, which was bare of ice. Rustling in the trees and walking under their canopies were mammals. They had been creeping by while the dinosaurs had lived, and by the Paleocene some of the major branches alive today had just sprouted. Yet when the dinosaurs disappeared, it was almost as if mammals were afraid of enjoying their good luck. Little dim-brained primates that looked like bug-eyed rats fed on insects and the fruit that the now-common flowering plants offered. Primitive herbivores, cow-sized at most, chewed on leaves and shoots—grass would not grow for another 30 million years. When the dinosaurs had lived, the browse line on the trees had been forty feet high. Now it was six.

One of the geological highlights of that time was India's collision with Asia. The Himalayas spiked upward, Tibet was pushed up into a plateau, and the Tethys Sea began to be crimped shut. It became like the Persian Gulf today, shallow, salty, and teeming with life. On its edges a new gathering of mammals appeared. Perhaps they had evolved in India during its missing years at sea, or perhaps they came from elsewhere in Asia. In either case, they were there in huge numbers: rodents in the thickets, bats and primates in the trees, and in the forests and scrub some of the earliest hoofed mammals—ancestral relatives of tapirs, rhinos, and elephants.

For predators, there were bobcatlike mammals called hyaenodontids, and there were the misfit mesonychids. The mesonychids came from gentle, herbivorous hoofed stock, their closest relatives probably the artiodactyls. They had a carnivore's taste for meat, but they had a horse's stiff back, which left them unable to run down prey. There was enough food to be had, though, using their bone-cracking jaws to scavenge corpses or sneaking up on an occasional turtle or splashing in the shallows of a river to catch a fish—not pawing it like a bear but snatching it in their long snouts. For the most part mesonychids stuck to this kind of life, raising hundreds of thousands of generations of young mesonychids, until 34 million years ago, when their particular brand of anatomical confusion could survive no longer.

Around the Tethys, however, one stock of mesonychids or some close relatives was drawn to the water. The sea was rich with herring and other fishes, and the salty water thrust itself up briny estuaries toward the interior. Like most mammals, these mesonychids could paddle across streams with the alternating stroke they used to walk on land. They had never used their hoofed feet to bring down prey; instead their long snouts became longer, so that they could snatch at fish farther away. Their front teeth turned into slender pincers that could hook swimming prey; with their bladed rear teeth they could cut the food, or they could simply swallow it whole. They fished much of the time in the shallows where water lapped up to their knees. When they swam they used up far more energy than when they ran. The water drained their heat, even though they lived in the tropics and had coats of fur. They had become *Pakicetus* and its peers—in other words, they had barely become whales.

Gingerich thinks that only 2 million years separate *Pakicetus* from the first, still-undiscovered true whale. Perhaps 3 million more years passed before *Ambulocetus* arose—Hans Thewissen isn't sure yet of its precise age. Evolution worked quickly in this time. New species of whales swam into deeper, wider water, maturing from a dog paddle to a yapok paddle. Their legs grew short, their hind feet long and probably webbed. Following Fish's model, it's likely that their fur, now buoyant, helped to keep their bodies straight as they swam. New species arose that could swim like otters, switching from surface paddling to a galloping kick, flexing their now loosening vertebral column, that could drive them down into deep river water. Why should whales bother to make such a transition? *Pakicetus* and its contemporaries did well only visiting the water—in some Eocene sites in Pakistan their teeth are now the most common fossils found. A coincidence may be a clue: at the same places where paleontologists find early whales, they find anthracobunids, those ancestors of elephants and manatees, followed in time by the earliest sirenians. Perhaps sirenians were the sleepy pioneers that went into the water first. They grazed like hippos underwater on grass at the bottom of rivers and coastal salt marshes, and whales, a lineage of predators that had already been flirting with the water, followed them in.

Yet *Ambulocetus* had not yet followed them in all that far, according to Thewissen. In his mind, they were furry crocodiles. They lay on the shores much of the time, their heavy heads resting on the sand or some high rock, and it may have been in this position—not underwater—that whale hearing got its start. Low-frequency sounds can travel rapidly through the ground, and some vertebrates can detect it. A sitting turtle, for example, can sense earthbound vibrations when they flow into the underside of its shell because the bone and the earth are acoustically similar. The vibrations travel through its skeleton to its skull, where they make its stapes vibrate against the cochlea. As *Ambulocetus* basked in the Pakistan sun with its massive head on the ground, these sound waves could have traveled up its bony jaw and channeled through its pad of fat to its ears. If Thewissen can confirm his speculation, whale ears may turn out to be like tetrapod legs: an exaptation so exquisitely well suited for one kind of environment that it is hard to believe that it originally evolved for life in another.

Like crocodiles, *Ambulocetus* was a different animal altogether when it slid into the water to hunt. It could ambush an animal by paddling slowly, its eyes riding high on its head as the rest of its body remained hidden underwater, until it reached a close distance and could break into an otterlike stroke. By seizing the animal in its long snout, *Ambulocetus* kept the desperate bites and claws of its prey at a safe distance. Crocodiles, like whales, can bind huge amounts of oxygen in their blood and muscles, but instead of using it to dive in the open ocean they use it to hold their breath while they drown prey in deep water. Ancestors of *Ambulocetus* may have evolved their capacity for oxygen for the same reasons. And once its prey was dead, *Ambulocetus* could use the powerful jaw and neck muscles evolved by its mesonychid ancestors to grab hunks of the creature's flesh and wrench it free.

It's tempting to build this story like a totem pole, with trotting *Pakicetus* at the base, *Ambulocetus* laying its humming jaw on top of it, and *Rodhocetus*, the earliest whale to swim like a whale, sitting above the two. It seems like such a smooth progression toward today's cetaceans that it must be right. But such a version would only be a vertical slice of the story. Life doesn't proceed from one point to another—like the cladograms that represent it, it forks and radiates. Thewissen and other paleontologists have found many other whale bones in Eocene rocks of Pakistan and India. Mostly they are teeth—the rock surrenders a few skulls as well—but even teeth clearly show that their owners were not clones of *Pakicetus* or the other better-known whales. *Ambulocetus* kept to brackish deltas and coastal water, but Thewissen has found whale teeth from about the same age in what at the time was the open ocean. Gingerich has found at least three contemporaries of *Rodhocetus* a few million years younger than *Ambulocetus: Takracetus*, with a wide, flat head; *Gaviocetus*, with a slender skull and loose hips; and *Dalanistes*, a whale with a head as long and narrow as a heron's set on a long neck, with hips cemented firmly enough to its spine to walk on land.

If this is a confusing picture, it should be. As time passed, certain whale species emerged that were more and more adapted to life in the water, but other species simultaneously branched away in many directions. Walking and swimming whales lived side by side, or in some cases traded homes as the buckling birth of the Himalayas shuffled their habitats. Some were only a minor variation on a theme that would carry through to modern whales, but others—heron-headed *Dalanistes*, for example—belonged to strange branches unlike anything alive today.

It's possible to track the shifting habitat of whales by looking at the rocks that hold their fossils—the grooved siltstone of flood plains of the most primitive forms, the rippled sandstones of a beach of more advanced species, and finally the blank limestone farther out to sea. But paleontologists have also figured out how to inspect the kidneys of extinct whales. The terrestrial ancestors of archaeocetes had spent 300 million years adapting their kidneys to surviving on land, using them to concentrate wastes to hold back their water. They were not ready to live in salt water. Manatees and other sirenians have been grazing along coasts for 50 million years, and in some ways they're still not ready: they have to drink fresh water occasionally to survive. Run a hose of fresh water overboard off the coast of Florida and they will pay you a visit. Whales, on the other hand, get enough fresh water from the air they breathe and the prey they catch, but also occasionally swallow salt water.

Oxygen in a river weighs less than oxygen in the ocean. The atoms that help make up water molecules in seawater tend to carry one more neutron than in freshwater, thanks to the physics of rain and evaporation (each kind of atom is called an isotope). When a growing animal drinks, oxygen in the water works its way into the developing bones and teeth, and so it was that Thewissen and a crew of geochemists and paleontologists were able to watch whales go to the ocean by analyzing the oxygen in their fossils. They first measured the isotopes of oxygen in the teeth of living cetaceans and found that river dolphins, which live in freshwater, had significantly lighter oxygen isotopes than marine species such as killer whales and bottlenose dolphins. Thewissen and the other paleontologists then surrendered ten teeth of early whales, including *Pakicetus* and *Ambulocetus*, to the geochemists. *Pakicetus* fell in with the river dolphins, as did *Ambulocetus*, but later species had teeth as isotopically heavy as a sperm whale.

Dalanistes

What's most interesting here is *Ambulocetus:* the rocks demonstrate that it swam in salt water either just offshore or in brackish deltas, and yet the water didn't leave its mark in the whale's bones. Perhaps it was like a manatee, drinking only freshwater, or perhaps it was more like a seal, nursing on land for months and entering the ocean only after growing fairly mature.

Between about 43 and 40 million years ago, the first species of whales probably arose that could for the first time do without a touch of land whatsoever. Although they still measured no longer than a horse and carried hind legs complete with toes, their kidneys now fought off the salty Tethys water and their heads were lengthening, so that their nasal openings were now no longer at the tip of the nose but halfway toward the eyes. Along with the early sirenians, these were probably the first mammals to give birth to a breathing baby underwater. They explored the edges of the Tethys, visiting Egypt and South Carolina. Instead of ambushing fish and sirenians in this tropical sea, they pursued their prey, swimming quickly thanks in part to their sleekly shaped bodies as well as to their muscles. The muscles that had once powered their limbs dwindled, while the ones that had merely supported the back and midriff expanded. In place of hair, they evolved blubber that made them as buoyant as water itself. The more archaic forms, the *Pakicetus-* and *Ambulocetus*-like species, all died away over time so that only the fully aquatic whales were left on earth.

By 40 million years ago one of these lineages of whales had given rise to true cosmopolitans. For the first time cetaceans broke out of the Tethys Sea, swimming to the cold North Atlantic, south to the Ivory Coast, around Cape Horn, up to Baja, and on to New Zealand. These whales, known collectively as basilosaurids, reached great sizes for the first time: they ranged up to fifty feet long, finally taking the place of the reptilian giants that had died 25 million years earlier. Among these new whales was *Basilosaurus,* the animal that had fooled people in 1832 and that Richard Owen had tried to rebaptize. Long after it was accepted as a whale, *Basilosaurus* showed up in paintings as curvy as a corrugated tin roof. It probably was nowhere near as exotic, instead looking something like an oversized version of a northern right whale dolphin, a fast, slim cetacean whose body has a similar ratio of length to width. Yet the skeleton of *Basilosaurus* shows that it was nevertheless an odd whale. All cetaceans alive today, for example, have squat vertebrae, but the spine of *Basilosaurus* was a chain of long, barrel-shaped bones. When paleontologists put clues such as these into their cladistic programs, they find that *Basilosaurus* ends up on a branch of its own and is therefore probably not very close to the lineage that gave rise to the whales on earth today.

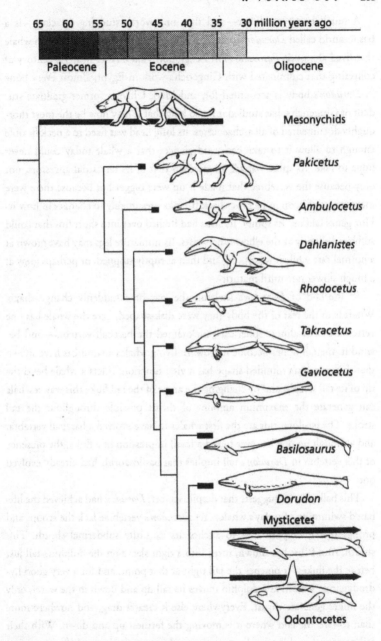

A provisional phylogeny of whales. While there are many species not shown, this tree includes the major branches of whale evolution. A few caveats: some reconstructions (such as *Pakicetus*) are based on very limited fossils. Also, because mesonychid relationships are so unclear, future research may show that whales actually have a more recent common ancestry with a particular mesonychid species. Finally, this tree is based only on morphology: molecular studies don't agree on some of the branchings.

A much better candidate—and thus one worth studying up close—is a basilosaurid called *Dorudon atrox*. This is the fifteen-foot-long, three-ton whale that lived alongside *Basilosaurus* in Zeuglodon Valley. After nearly a century of collecting that culminated with Gingerich's work in Egypt, almost every bone in *Dorudon*'s body is accounted for, and Mark Uhen, a former graduate student of Gingerich's, has studied it for so long that it may now be the most thoroughly documented of all archaeocetes. Its long head was fixed to a neck flexible enough to allow it to take backward glances that a whale today could never hope to take. Its spine was far longer than that of its terrestrial ancestors, not only because the vertebrae that made it up were bigger but because there were many more of them in its lower back (thanks presumably to changes in how its Hox genes laid out its spine). Its arms had fleshed over into short fins that could still bend slightly at the elbows and wrists. Its miniature legs may have grown at a normal rate while it was young and then abruptly stopped, or perhaps grew at a much slower rate until maturity.

At the end of *Dorudon*'s skeleton the vertebrae suddenly change shape. Whereas in the rest of the body they were disk-shaped, here the whale has one vertebra that is almost spherical—nicknamed the baseball vertebra—and beyond it, the tail bones become narrow. In living whales anatomists have shown that this vertebra's rounded shape has a vital function: it lets a whale bend the tip of its tail steeply, and by changing the angle of the tail fluke this way, a whale can generate the maximum amount of thrust possible throughout the tail stroke. The basilosaurids are the first whales to have evolved a baseball vertebra, and although no one has ever found a fossil impression of a fluke, the presence of this vertebra in *Dorudon*'s tail implies that basilosaurids had already evolved one.

This ball of bone suggests that despite its feet, *Dorudon* had achieved the lift-based swimming of today's whales. Yet *Dorudon*'s vertebrae lack the scoops and prongs that a dolphin uses to anchor its exquisite subdermal sheath. This sheath, Ann Pabst has shown, turns into a tight sleeve on the dolphin's tail just before the flukes. It pinches the tail tight at that point, and for a very good hydrodynamic reason: as a dolphin moves its tail up and down in the water, only the flukes generate thrust. Everywhere else it creates drag, and nowhere more than toward the end where it is moving the farthest up and down. With their tail narrowed at this spot, living cetaceans become much more efficient swimmers. *Dorudon* lacked this sleeve, suggesting that its tail probably had the wide shape of a manatee's. The sheath must have evolved in some later cetacean, growing from a small stretch of connective tissues that many mammals have in the lower back to a full-body sleeve.

Dorudon might have been a decent swimmer, sprinting twenty-five miles an hour, but without the sheath it was far less efficient than a dolphin. Its wide tail created drag, and it couldn't reclaim any of the energy of a kick in a springy sheath. It had to work hard to swim, and its blubber threatened it with overheating. *Dorudon* or one of its immediate ancestors most likely solved this problem by seizing on the patterns of blood vessels that had served its ancestors on land. Mesonychids probably had kept themselves cool in the tropics by dumping heat into the skin of their long tail and limbs. Whales reengineered their blood vessels so as to restrict the flow of heat only to their flukes and fins. And with its veins rerouted so that they flowed to its testicles and womb, bathing them in cool blood, *Dorudon* escaped the danger of sterilizing itself with heat.

A hint of what *Dorudon* ate comes from its cousin *Basilosaurus:* one specimen was dug out of the Yazoo clay of Mississippi with a muskmelon of stone where its stomach once was. Inside were teeth and bones from herring, two-foot-long sharks, needlefish, and sea robins, all etched by stomach acid. Clearly whales had given up crushing bones and turtle shells for millions of years; relieved of these stresses, the lower jaw became lighter, hollowed out and filled with fat that could funnel underwater sounds to the ears. Without echolocation, *Dorudon*'s ears, which were tuned to low-frequency sounds, could probably only pick up the sounds of other fish passively. It pursued them individually, snagged with its slender front teeth, and sliced them apart with jagged molars.

Dorudon was a hefty whale that swam many miles from shore, an animal that would die if it was beached, and yet it does not fit our notion of a whale. It bore legs and toes, its head was small, its neck was relatively long and flexible, its flippers could bend like arms. It couldn't echolocate like toothed whales or bulldoze through krill like a baleen whale. It had a small brain, and its swimming left a lot to be desired, hydrodynamically speaking. Although *Dorudon* and the rest of the basilosaurids had millions of good years in the ocean, they hadn't achieved the modern forms that living whales take. Much of what we think of as the successful adaptations whales made to the water were still to come: the transition took far longer than the actual plunge.

Toward the end of the Eocene, 34 million years ago, the world began to cool again. Forests pulled back from savannas and the ocean's currents shifted. Like the walking whales that had vanished long before, the basilosaurids disappeared. *Basilosaurus*, the first archaeocete ever found by humans, was among the last to die. The reasons still aren't clear. Perhaps these early whales were simply tropical at heart and faded when the waters became chillier. On the other hand, by the time they were dying off, paleontologists think, new kinds of whales had appeared.

The origins of baleen whales and toothed whales have been almost as hard to tease out as the origin of whales altogether. Even ten years ago it was still possible for a few people to look at the evidence at hand and reasonably suggest that archaeocetes, mysticetes, and odontocetes descended from three different terrestrial mammals. While archaeocetes share some features in common with baleen whales and toothed whales, they have such distinctive bodies that a compelling connection was hard to find. Archaeocetes and odontocetes both have teeth, but the teeth of archaeocetes are differentiated dental work—molars, premolars, and so on—while most toothed whales have jaws full of identical pegs. What makes odontocetes most distinctive are the many ways they are adapted for echolocation. A melon or a monkey lip can't fossilize, but the air spaces and reflecting skull bones that accompany them do, and archaeocete skulls lack them all. These early whales could hear reasonably well underwater, but they couldn't cast a sonic flashlight through the ocean.

Baleen whales seemed even more isolated. Contemplate a blue whale's tongue—which alone weighs as much as an elephant—or its pleated lower jaw that can swing open to engulf seventy tons of water, or its forest of baleen plates, and you'll find it easy to appreciate the paleontologist's problem. A possible solution appeared in 1966, cloaked in a whale skull that tumbled out of the wrong time. Found in Oregon, it belonged to a whale that had lived late in the Oligocene period (34 to 24 million years ago). Although the whale, named *Aetiocetus*, lived millions of years after scientists thought the last archaeocete had died, it didn't qualify as an odontocete (its skull wasn't modified for echolocation) or a mysticete (it had half-inch-long teeth instead of baleen). For a time *Aetiocetus* was treated like a freakish, latter-day archaeocete. But soon afterward Leigh Van Valen argued that this label was wrong: *Aetiocetus* qualified in almost every way as a baleen whale, he pointed out, except for the baleen. Its lower jaw bones were loose at the chin, and the upper jaw formed the same kind of buttress underneath its eyes that living baleen whales have. "The ancestors of mysticetes must have had teeth," Van Valen observed, "and *Aetiocetus* is in other respects similar to mysticetes." It was, he decreed, a baleen whale with teeth instead of baleen.

With such scant evidence, Van Valen could make little more than a rough guess, but since the 1960s paleontologists have come to agree that he was right. They have recognized a number of other baleen whales with teeth, some as old as 34 million years, soon after the last archaeocetes died. There are five living families of toothless mysticetes, but paleontologists have identified three extinct families with teeth in the past few years. One of the most recently discovered of these fossils shows particularly well how the transition might have happened. Its discoverer, Larry Barnes of the Natural History Museum of Los Angeles County, has spent a fair portion of his career demonstrating that Van Valen was right about baleen whales. He has traveled to Japan, Mexico, South Carolina, and Oregon, and has found toothed mysticetes in all these places. He has actually dug only a few of them out of the ground. "What really has been fun is that I've found them in collections that other institutions have held, where these things have been identified as archaeocetes or not identified at all. And I've said, 'Wow, do you know what you've got?' And they've said, 'Uh, no.' "

In 1996 Barnes had one such experience at the South Carolina Museum of Natural History. He realized that three different species of whales belonged to a completely new family of mysticetes. As 25-million-year-old creatures, they aren't the oldest baleen whales, but they are the most primitive Barnes has ever seen. He suspects that in their day they were living fossils that preserved much of the anatomy of the first baleen whales. He gave them a suitably primordial name: *Archaeomysticetus*—ancient baleen whale.

Archaeomysticetus was small by the standards of today's baleens: its five-and-a-half-foot skull implies a thirty-foot body. It had a loose lower jaw, and the upper jaw was beginning to sweep back behind its eyes, carrying the blowhole with it. Yet its skull lacked the pockets of air that odontocetes use for echolocation. "It has all those structures of a baleen whale, but its teeth could be taken out of a *Basilosaurus* skull," says Barnes. They are spaced out in the jaw as in *Basilosaurus*, they have the same serrations, and most important, they are the same size. Barnes has mostly encountered puny, human-scale teeth in primitive baleen whales, but *Archaeomysticetus* had teeth four inches long and three inches wide. You don't cup them in your palm; you try to wrap your fingers around them.

Although it was a baleen whale, *Archaeomysticetus* was not like today's versions. It still hunted like its older relatives, chasing South Carolina sharks and swallowing them whole. In fact you could easily turn a basilosaurid into *Archaeomysticetus* simply by extending the upper jaw underneath its eye and loosening the fusion between the sides of its lower jaw. "And that's pretty easy to do," say Barnes. "That's not much evolutionarily. Dolphins and porpoises—there are bigger differences between them. If whale evolution stopped at *Archaemysticetus,* we wouldn't say they were baleen whales, because we wouldn't have seen anything with baleen plates. Whenever you're looking at a transitional stage, the differences are minor. Only from the perspective of where they have come to, after another 30 million years of evolution, do we appreciate what those differences were. Because of our perspective, we draw the line back, and this is where we'll cut it off."

Other fossils suggest a further evolution toward true baleen whales. The nose moves all the way toward the top of the head. The teeth become smaller and uniform, and as they do so they multiply to cram the jaws. The embryos of today's baleen whales still retain this pattern of teeth buds, but they soon vanish under a curtain of baleen. How baleen itself came to be remains a gap to fill. It rarely fossilizes—the oldest baleen plates are 15 million years old—but judging from the jaw structure of older whales, baleen probably evolved much earlier. A living cetacean could be a model for this final step in the transition to baleen whales: Dall's porpoise, a six-foot-long, white-flanked cetacean that lives in the North Pacific, is a toothed whale, but its minuscule teeth are completely surrounded by a horny set of gums that it uses to grip squid and fish. On a microscopic scale, this hard tissue is almost identical to baleen. Perhaps some early toothed mysticete 30 million years ago might have followed a similar path, developing gums as tough as nails for catching some particular prey.

The first true baleen whales may have been the product of a rapidly changing world. The continents were moving closer to their geography today, with Antarctica isolated from the other continents. By the Oligocene it was squarely over the South Pole and the ocean was able to rush full-circle around its edges. Cordoned off from the rest of the world, it grew cold and icy. Chilled, dense water dropped down to the bottom of the ocean and flowed toward the equator, rising back up to the surface along continental shelves. Plankton thrived where it rose, carrying nutrients from the deep, and their dense swarms made it possible for a big predator to scoop them up if its anatomy allowed. Some gummy baleen whales shifted from using their mouths to grab large fish to swallowing clouds of these small creatures and then squeezing the water out to sieve their food.

Evolution was able to work on many characters from their past to make them better filter feeders. Their jaws, unfused, could swing wide open to gulp and capture more food. The broad shelf of the upper jaw grew stiff enough to keep the lower jaw from dislocating as it swung open. Evolution had never created such an effective way to shovel up the ocean's wealth, and the bigger whales grew, the more food they could gather, until some lineages reached the sizes we see today. With no predators to worry about and no fixed territory to protect, the baleen whales gained the loose, casual societies they have today. According to this scenario, baleen whales never evolved any detectable level of echolocation because they never needed it, nor could they even try once their skulls had been so altered for filter feeding.

The names biologists have given whales are turning out to be deceptive. Baleen whales apparently arose before there was baleen, and although toothed whales all have teeth, archaeocetes had them as well. To find truly unique defining features of toothed whales, you must look to their melon, monkey lips, and other echolocating organs. The origin of these features is, for all intents and purposes, the origin of toothed whales, but their fossils don't help nearly as much as baleen whales fossils do. The oldest toothed whale fossil, found in Washington State, dates back 34 million years. "It's got everything it takes to be an odontocete," explains Barnes. "It's got the sacs where the air sacs went, it's got the maxillary bone where it needs to be for the squeezing of the melon." The echolocation system starts, at least so far as the fossils can suggest, out of the blue.

By comparing living toothed whales with fossils of archaeocetes, though, some researchers have at least been able to put some boundaries to the possible paths that evolution could have taken. The origin of echolocation probably depended on two of the most common features of macroevolution—exaptations that the ancestors of odontocetes already used for some other function, and the correlated progression of many different parts. The ears of toothed whales, having been partially isolated from their skulls for 10 million years, were already protected from their own clicks and were thus prepared for echolocation. To make the actual cries, odontocetes must have evolved the ability to sing through their noses instead of their voice boxes. Their common heritage with artiodactyls may have helped here: mammalogists have noticed that ibex, chamois, and gazelles all make alarm calls through their noses. Perhaps mesonychids could as well.

Their melon may have already existed as a nose plug. In dissections of baleen whales, John Heyning of the Natural History Museum of Los Angeles County and James Mead of the Smithsonian Institution have found small blobs of fat and connective tissue near the blowholes that look like miniature melons. They suggest that melons may have begun as nose plugs for the first archaeocetes to dive deep. These early whales would have needed to clamp their blowholes firmly to keep salt water from getting into their nasal passages. Like living whales, basilosaurids had shelves of bone on the top of their heads, which they used for the same purpose—to anchor muscles that slammed the blowhole shut. In order for these muscles to glide smoothly, the whales might have evolved a fatty structure surrounding the blowhole to lubricate the path. In mysticetes we still see this ancestral pad, the argument goes, but in odontocetes it swelled and took on a new function.

Here was raw material that evolution might have used to produce echolocation. It could not have developed any one part of the system to the exclusion of the others—what point would there be in a whale becoming able to hear high-frequency sounds if it couldn't produce them in the first place? But if changes happened in increments all over the heads of toothed whales, each would encourage the other. Perhaps when some whales accidentally made a noise in their nose, they could faintly make out echoes of a neighboring fish, giving them a slight advantage in hunting. Sound might have been inadvertently focused by their nose plugs, and so whales born with oversized ones might have been favored. Meanwhile, the nose was moving up toward the top of the head for what anatomists suspect was an entirely unrelated reason: to make breathing more efficient. But in order for the nose to make its trip, the bones of the upper jaw had to expand back toward the eyes to carry it there. The farther back the upper jaw went, the more stable the whales' skulls became, which helped them in hunting. The same transformation, however, created a reflecting dish on the upper jaw for sound waves coming from the nose, as well as a platform on which the melon could rest. The ears nudged their way up to higher and higher frequencies as they separated even farther from the skull. This possible web of changes carried on for thousands of generations until echolocation, having started as a minor extra clue to the whereabouts of a fish, lit up the ocean.

Molecular biologists discovered the structure of DNA in 1953, but only in the past decade have they been able to easily read actual sequences, base by base. They have read the genomes of strains of bacteria and fungus from beginning to end, and the Internet creaks with databases of their four-letter codes, reading like the yearlong tappings of a monkey on a four-key typewriter. When news of freshly discovered genes turns up in the newspapers, the story usually involves grim enlightenment (this gene or that hikes up your chance of this cancer or that) or tenuous correlation (people with a certain gene tend to avoid rock climbing). At the same time, though, some scientists have been using newly sequenced genes to quietly create a revolution in evolutionary biology. They have been comparing the genes of different organisms and constructing cladograms. Like Geoffroy and Owen searching for homologies in the shape and texture of bones and glands and teeth, molecular biologists can look at the sequences of genes for homologies of code. And just as you can turn touchable homologies into cladograms, you can use the ones read by a gene sequencer. Say that in one four-position sequence of an ancestor's DNA, the code ran ACGT. As its descendants branched into species, some of the positions might mutate while some would remain the same. A gene tree would then look something like the one shown here.

The craft of building gene trees (known as molecular phylogeny) is already changing the study of macroevolution for good. Where anatomy confuses, where fossils are nowhere to be found, molecules shine a light. At the moment the discipline is brash and young, and its pioneers are still figuring out what they can and cannot learn from their data. The molecular phylogeny of whales is an important example, because the version of whale evolution that the bones tell so eloquently doesn't agree on some key points with some of their genes.

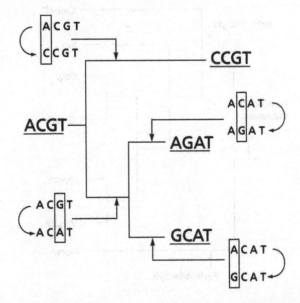

Because an ancestral gene sequence mutates in descendant species, you can use it to build a cladogram.

Several teams of geneticists have been comparing the genes of whales to those of other mammals since 1991, and they've been converging on a startling pattern. A simplified version of one of the latest trees, created by a biologist at the University of Arizona named John Gatesy, is shown below.

This tree says something very different from the old work with rabbit blood. It claims that whales not only are relatives of artiodactyls, but are actually artiodactyls themselves and have as their closest living relative the hippo. It would make the lives of many paleontologists much easier if this tree didn't exist. Many find it hard to square with their evidence that mesonychids gave rise to whales. All artiodactyls have a unique double-pulley bone in their ankle, but mesonychids do not. This fact, along with many others, makes paleontologists conclude that mesonychids aren't artiodactyls. And if whales are more closely related to mesonychids than any other animal, then they cannot be artiodactyls either.

If Gatesy's tree is true, the paleontologists are left with two equally painful choices. One is that mesonychids actually were artiodactyls closely related to hippos and lost the double-pulley ankle—something no other artiodactyl has been foolish enough to do. The other choice is that mesonychids are ungulates but not artiodactyls, and their many whalelike traits are all a convergent illu-

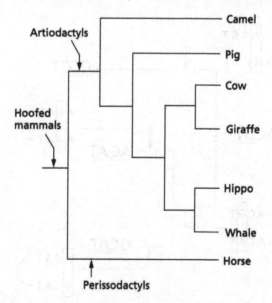

Some research on genes suggests that whales are artiodactyls closely related to hippos.

sion. There was a time when paleontologists in Pakistan picked up loose mesonychid teeth and wondered what new species to call them. Years later they found the same teeth attached to skeletons and realized that they were actually whales. Given how distinctive teeth are, many paleontologists are loath to accept that whales and mesonychids evolved such similar ones independently.

As for the hippos, Gatesy points out that like whales, they are hairless, water-loving animals that can nurse their young underwater. According to the fossils, hippos, which date back 15 to 20 million years, probably descend from a terrier-sized stock of artiodactyls whose fossils in turn reach back perhaps 40 million years. Paleontologists also concluded that both hippos and these ancestors are on the pig branch of the artiodactyl tree. The new gene work would tear hippos and their ancestors out of this group. It also creates a gap in the fossil record: if hippos and whales share a close common ancestor, then it must be older than *Pakicetus*, which is 49 million years old.

However whales came to be in the first place, though, the fossils now at least tell a reasonably smooth tale of how they evolved into the major groups alive today. And here too a geneticist has made life interesting. In the early 1990s Michel Milinkovitch, now at the University of Brussels, decided to use some new methods for sequencing DNA to see how whales are related to one another. Friends in the marine mammal business sent him DNA, and he got hold of sequences from artiodactyls against which he could compare it. The tree that he found, and that some subsequent work on more genes supported, is shown on this page.

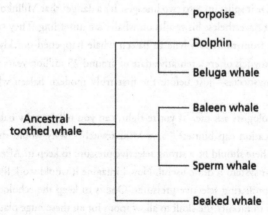

Genes also suggest that baleen whales are actually descendants of odontocetes, and that their closest relatives are sperm whales.

Milinkovitch was sure that the tree was wrong. "It was so much in my mind, as I guess it was in the mind of everyone, that toothed whales are a mono-phyletic group." In other words, he assumed that the animals that William Flower named odontocetes were in fact descended from a common ancestor that begat no other group of animals. Perhaps that sentence is lost in the ether of taxonomy. Here is a more concrete way to think of why the tree gives pale-ontologists conniptions: They envision baleen whales arising 35 million years ago or earlier from a primitive basilosaurid with big teeth and no echolocation. Toothed whales arose from a separate lineage, whittling their teeth down to a se-ries of pegs or a single giant beak and engineering their echolocation system. But according to Milinkovitch's cladogram, when the archaeocetes died out, only odontocetes carried on the genes of whales. Millions of years later a lineage of odontocetes closely related to sperm whales abandoned sonar hunting and dismantled all the elaborate equipment necessary for the signals. What anatomists have identified as a nose plug in baleen whales is actually a vestigial melon. (The alternative—that sperm whales and non–sperm whales indepen-dently evolved sonar—seems less likely to Milinkovitch.) Mysticetes quickly evolved baleen to scoop up hordes of shrimp and other small fry.

Since his original study, Milinkovitch has used other genes and new kinds of analysis, but the results are always essentially the same. Milinkovitch still won-ders, though, if his results are flawed. Gene sequences don't necessarily change and diverge in a way that precisely matches the branching of species. It's possi-ble, for instance, for a gene to evolve into two or more versions within a single species, *before* it splits up into two lineages. It's a danger that Milinkovitch is ex-ploring, but nevertheless his results on whales are unsettling. They suggest that the change from toothed whale to baleen whale happened quickly; from his data Milinkovitch offers a tentative date of around 25 million years ago for the origin of mysticetes—just before the first truly modern baleen whale fossils appear.

"Morphologists ask me, 'If you're right, can you tell me why baleen whales lost echolocation capabilities?'" says Milinkovitch. "Why would they do that given that there should be a strong selective pressure to keep it? After all, a bio-logical sonar system is quite useful. Now I imagine it would work like this: you have two conflicting selective pressures. One is to keep the echolocation, and the other is to modify the skull to allow room for all these huge plates for filter feeding. It looks likely to me that there was conflict, because having this large flat upper jaw and this large curved lower jaw does not let echolocation work properly anymore. Now, if you have a group of organisms starting to filter-feed, this is a brand-new way for mammals to feed in the sea. Having the possibility to occupy a new niche wins, especially because echolocation is not that important when one is looking for huge fish schools. That also would explain how baleen whales so quickly modified their morphology. Baleen whales radiated suddenly because they found a new way of feeding. That's how I see it."

Milinkovitch is a cheerfully blunt person to talk to. He lashes out at people for what he sees as sloppy science, calls their papers stupid or mean, and yet seems a bit baffled at how some of them don't want to talk to him afterward. The backlash he has received has taken the form of a retelling of the story of a few pages back, of toothed baleen whales grading smoothly from archaeocete to baleen whale. He is not moved. "What I would like to get—and what I haven't gotten yet although I asked several morphologists—please, send me a matrix with all the characters and let's do a cladistic analysis with your data. Nobody sends that to me. No, they prefer to show a skull and then another skull and then another skull. And then they say, 'Look, there is sort of a smooth path through these different skulls.' And that's weird. Many of these could be side branches that have nothing to do with the origin of baleen whales or sperm whales or any living whales. These are extinct branches. Another problem is that a lot of these specimens are incredibly incomplete. They show you a drawing, but you should see the fossil!" Several of the earliest fossils labeled as toothed baleen whales are known so far from single teeth, with no whale attached. "Of course they might have interesting characters, but it's very different from having very complete fossils like Thewissen and Gingerich have found. These are good fossils for sure."

No one, it is true, has yet published a cladogram of all the fossil creatures that paleontologists think are on the line from archaeocetes to today's baleen whales. Part of the problem lies in the fact that many of the recently discovered fossils turned up so suddenly that their discoverers have had to stumble to conferences with carousels full of slides so new they haven't seen the pictures themselves. Getting the raw data to make a molecular tree is mainly a question of firing up gene sequencers and computers; paleontologists have to eyeball fossils—not only the ones they dig up themselves but those in museums around the world—and decide what crushed skulls looked like in life, whether a certain rib was absent or is still waiting to be found in another rock somewhere. But what would Milinkovitch do if the smooth story survived the metamorphosis into a real tree?

"Then it becomes really exciting," says Milinkovitch. "Let's say they did it really well and didn't include the dubious characters and the two trees disagree. Then of course one of these must be wrong. I'm not trying to say that the molecular one will necessarily be best, but at least I have some reasons to believe that molecular data will be more informative. At least *we* know what a character is. If you take position thirty-two in the cytochrome b gene, that's a character. That's a fact. If you get an *A* there for a baleen whale, that's unambiguous. Now take the same thing for the morphology: what is the character? Well, that bone, or that part of the bone. Or maybe a part of a part of the bone. That's an interpretation—it's subjective.

"Another problem is more conceptually major. You don't inherit a morphological character from your father and mother. You build it anew. That's also a fact. Obviously that has huge consequences, because between the zygote and the fully grown adult individual many things happen during development." The crest on a bird's shin is not encoded anywhere in the genes but emerges as it fidgets in its egg. A structure shared by two species, like a jaw, may look homologous to a morphologist, but in some cases different animals build it with different populations of cells. The debate over the evolution of whales is, at its heart, about much more: it's an arena where biologists can struggle over the deep questions that Cuvier and Geoffroy and Owen grappled with years before *On the Origin of Species*—what is homology, and how can we recognize it?

Anatomists and paleontologists are happy to leap down from the stands to join the molecular phylogeneticists. After speaking with Milinkovitch I grabbed the Sixth Avenue subway uptown to the American Museum of Natural History again, where I spent a morning with Patrick Luckett. Luckett is an anatomist and mammalogist from the University of Puerto Rico. He specializes in comparing the reproductive and developmental structures of mammals, and visits the museum's collection for at least a month every year. He is also an editor of the *Journal of Mammalian Evolution*, a new quarterly that tries to reconcile molecules and morphology. "I'm very much in favor of molecules, and I'm trying to bring the two together," he says.

But Luckett also thinks that molecular biologists need to think more like anatomists before they will all find a common ground. Luckett isn't vicious or doctrinaire—he is like a new grandfather, gentle and a little unsure of what to do with his wisdom, only a few spots of age on his hands. He extended one of them across his table, which looked as if it had just been caught in a hailstorm of skulls and teeth. He pulled one bone over to us, the skull of a baby pygmy hippopotamus. Pygmy hippos have the same eggplant-shaped bodies as their bigger, more familiar relatives, but they grow only five feet long. The name hippopotamus—"river horse"—is wasted on the pygmy hippo: it doesn't spend much time in water but wanders the forests and swamps of West Africa at night, eating fallen fruit and tender shoots. Regular hippopotami spend far more time underwater, and their bodies show it—their eyes, for instance, sit on top of their heads like the eyes of a crocodile or *Ambulocetus*. Likewise, the terrestrial life of a pygmy hippo is clear in its skull, which looks like a bear cub's, with eyes on the sides. Although these two species are the only hippos alive today, many other forms have existed. Some were amphibious, but others were even more terrestrial than the pygmy hippo. Hippos once climbed the Alps.

The idea that hippos—as cousins of whales—still hold onto the amphibious habits of the common ancestor of both animals may rest on some simplistic assumptions. The only traits we know with certainty that the common ancestor of all hippos had are the ones that we find on all living and fossil hippos—not the ones that one lineage of hippo or another evolved on its own. This fact guides Luckett when he builds a cladogram of any set of mammal families: he uses only the characteristics that each family's founder must have possessed. He drew my attention to one such trait as we sat at his table of bones. One of the baby molars on the lower jaw of the pygmy hippo looks like three teeth soldered together. Hippos share this trait with cows, camels, and all other artiodactyls, but no whale has it, nor any other mammal.

Luckett is happy to talk to molecular biologists on their own terms, and the pygmy hippo is important to him in that respect as well. Just as it's easy to assume that all hippos are alike morphologically, it's also possible to think that they have indistinguishable genes. Yet molecular phylogeneticists often discover that if they include the genes of two species from one family in their analysis, they come up with a different tree than if they use only one. That's because when a computer uses a single species as a representative for an entire family, it has no way of telling which part of the gene existed in the ancestor of the family and which evolved later in that particular species. The problem only grows worse when molecular phylogeneticists use the genes of wildly diverse groups of animals. No single rat can stand in for over two thousand species of rodents.

The work of Gatesy and Milinkovitch recently inspired Luckett to look at the molecular phylogeny of whales with the eye of an anatomist. Rather than dumping raw gene sequences into a computer and letting it sort out the best—albeit shaky—tree, he picks out the bases he will use for his study, just as he picks out teeth and bones. Often when he looks at a gene from several species from the same family, there are obvious, long sequences that are identical: these, it's safe to assume, were also present in their common ancestor. "Computers can't think," Luckett says. "I'm trying to reconstruct what's ancestral, to reduce a family to one code. One species of hippo might share a feature with one or two whales, but the ancestral whale didn't have it." Only after he selected these family-level sequences did Luckett then build a cladogram. When he did, the hippo-whale link disappeared, and Milinkovitch's supposedly clear-cut picture of sperm whales as the closest relatives of baleen whales dissolved as well. "They have some characters in common with mysticetes, but they have others in common with other odontocetes," he says. They even share some characters with sirenians—manatees and dugongs.

In this tie, Luckett lets morphology tilt him, particularly the traits such as the double-pulley ankle and the baby molar that are unique to one group of animals. And for Luckett, morphology for now still argues that whales are not artiodactyls, and that Flower's traditional split of the whales between the mysticetes and odontocetes was right. Luckett doesn't share Milinkovitch's faith in the homology of genes, because they can create a confusion of their own. "We may know that this position is a character, but what does it mean when it changes?" he asks. Molecular biologists are tormented by genetic palimpsests. A position in a gene may start as a *G*, switch to a *C*, a *T*, and back to a *G* again. When anatomy changes, its ancient form may leave faint scars behind, be it a fishlike aorta or ear bones that first form in an embryo still connected to the jaw. When a base in a gene changes at one position, no trace is left of its original identity or the intermediate versions. Once two lineages diverge from a common ancestor and build up more and more mutations, more of this overwriting takes place. When a molecular biologist tries to follow these branches back through time, this overwriting attracts them toward each other, and the wrong branches may join together, producing a false tree. Attracting branches are not the only problem with molecular phylogeny, as Luckett points out. "In some genes you get gaps, losses, additions, insertions. The problem becomes, how do you align the genes?" There may be homology in genes, but after they get shifted one way or another, it may become impossible to recover.

To be fair, molecular biologists are more aware than anyone else of these shortcomings, and most of their work goes into making up for them. They measure how quickly different stretches of a given gene mutate in order to know how much track-hiding and stranger-imitation to expect, and then they select the best genes available for the question at hand. When their computers spit out a tree, they can shake it with a number of tests—throwing in a fake base here and there in the sequences, or tossing out one of the animals, and then recalculating the tree. If the tree is strong, the branches will stay in place; if the tree is the product of a lot of deceptive convergence, the branches may flutter around.

As biologists get a sharper view of how genes mutate and how embryos develop, such conflicts between genes and morphology will gradually disappear. But even in its infancy, molecular phylogeny has offered a solid confirmation of Darwin's ideas. A dozen species can theoretically be arranged into millions of different evolutionary trees, but genes generally offer a few alternatives that resemble one other. The genes of whales tell us that they are not tigers gone to sea, or anteaters or wallabies or bandicoots. Their closest relatives alive are the artiodactyls, or perhaps one artiodactyl in particular. The reality of evolution is not at stake, but the particular path that it took is.

The whales of the Tethys 40 million years ago had all the mental graces of a pig—not that pigs are lacking in intelligence as far as ungulates go. But at some point after the archaeocetes became extinct, whale brains blossomed, evolving quickly to the point now where they rival human brains in some respects. Intelligence is difficult to measure or even fully define in any animal, human or otherwise, and understanding how macroevolution can produce it is even harder. Yet thanks in large part to the work of Harry Jerison of the University of California at Los Angeles, those who try to study its evolution at least have a starting point: the relative sizes of brains.

Jerison first had to redeem brain size from its dubious scientific history. Since the nineteenth century many scientists have been convinced that a large brain means an intelligent one, but they were steered wrong by prejudices and bad theory. Some were sure that Europeans had the largest brains of all humans, or that women had smaller brains than men, perhaps because the European men addressing the question assumed they were the most intelligent of all. Within a single species, though, brain size actually says little about intelligence because the variation from one individual to the next is too small. Among species it can be meaningful only if one measures the weight of a brain relative to the animal's body, because, as Jerison argued, a huge chunk of any brain has to be dedicated to the quotidian details of maintaining a body—things like contracting muscles, making bowels ripple, sensing pain on its skin. Magnify an animal and you have more territory to manage, more muscles to twitch, more skin to monitor—and thus you need more gray matter. The fact that a moose has a larger brain than a mouse in itself means nothing because its body weighs five thousand times more.

Subtract the housekeeping parts of the brain and what is left over is the only thing that matters, because these extra neurons can integrate information from the outside world into abstractions—an ability than can serve as a rough definition of intelligence. Jerison made a thorough study of the relative brain weight of vertebrates and found that fishes, amphibians, and most reptiles fell along one line rising along a graph, but birds, mammals, and a few dinosaurs fell along a much higher one. In other words, a mammal might weigh the same as a reptile but have a brain ten times larger. In the transition from synapsid to mammal our ancestors rose from the reptile ratio to the mammalian one.

Relative brain size is like a statistical knife for dissecting the brain, according to Jerison, but to slice his way to the differences among mammals, he needed a sharper blade. He calculated from his data how big a mammal brain "ought" to be—the weight of an average-brained mammal of a given weight—and then factored in how far above or below this average a particular mammal's brain actually was. From the numbers he calculated its encephalization quotient, or EQ. At an EQ of 1, a mammal (a horse, for example) has all the brains it needs to live a standard mammal life. If its EQ is higher, extra nerve cells are available, and they can usually be found bulking up the neocortex where the senses pump their data and the brain chews them over.

Jerison threw EQ out as a theoretical construct in the 1970s, a skeet to be shot down, and while a few marksmen have clipped it, it's still relatively intact. When animal behaviorists test closely related animal species for their ability to solve problems, their performance matches well with their EQ. But some researchers have pointed out that simply having a lot of extra cerebral transistors doesn't automatically produce intelligence. As Harvard anthropologist Terrence Deacon has noted, a chihuahua has a high EQ for a dog, but it's certainly not a canine genius. Intelligence may not be determined by relative brain size per se but by the rate at which the brain grows in the embryo. In the womb, primates grow with an unusual timing: their brains swell early and quickly, but their bodies grow slowly for mammals of their size.

The growing brain creates much of its structure in response to the body in which it finds itself. Nerve connections make their way from the trunk and limbs into the regions that will control their movement, eyes to the places where vision will be created. When an undersized body is finished making its claims on an oversized brain, there is a lot of open brain space still left over, and the regions of the brain that don't pipe in data from the rest of the body—including regions of the neocortex that confer intelligence—can grab it. While it's in the womb, a chihuahua's brain grows at a normal pace for a dog while this crucial divvying up takes place. Later its body's growth rate slows down, giving it a deceptively high EQ.

Few mammals have brains and bodies that grow along the same trajectories as those of primates, but it seems that cetaceans are among the ones that do, and these two groups of mammals are also the ones with the highest EQs. Consider the selection of rankings shown below from a recent survey by Lori Marino, with a few ancient hominids thrown in to make things interesting. *Australopithecus afarensis*, best known from the fossil Lucy, lived 3.2 million years ago and was close to the ancestor of all later hominid lineages. *Homo habilis* lived about 2 million years ago, and *Homo erectus* lived from 1.8 million to 50,000 years ago; our own descent probably passed through these two species.

By far, humans are the most encephalized living animals on the planet, with a brain seven times bigger than an average mammal our size. But you have to make your way down the list through a lot of cetaceans before you get to the next living ape. In fact, until 1.8 million years ago, dolphins had higher EQs than hominids. Before then, the most encephalized creature on earth lived in the ocean.

ENCEPHALIZATION QUOTIENTS	
Homo sapiens	7.06
Homo erectus	5.5
Tucuxi (a Brazilian river dolphin)	4.56
Pacific white-sided dolphin	4.55
Common dolphin	4.26
Bottlenose dolphin	4.14
Risso's dolphin	4.01
Homo habilis	4.00
Dall's porpoise	3.54
Australopithecus afarensis	3.00
Killer whale	2.57
Chimpanzee	2.34

Because we apes and cetaceans are alone at the top of the EQ scale and grow our big brains in the same way, the only way to understand how whales evolved their minds is to compare them to ourselves. Jerison first invented EQ in order to find a way to gauge the intelligence of extinct animals, and by measuring the empty braincases of 60-million-year-old primates he showed that these distant ancestors had rat-league EQs. Their skeletons show that they were lemurlike animals that could leap from tree to tree, and about 50 million years ago one lineage of their descendants became small diurnal clamberers. Their eyes grew more sensitive and moved forward to the front of their heads in order to make out insects more easily. From this stock the first apes arose 30 million years ago and lived comfortably in the trees and on the forest floor, shifting to a diet of mostly fruit for 20 million years. But 10 million years ago, with further cooling of the climate and shifts in vegetation, they were relegated to margins of the forests.

Evidence exists for several different evolutionary pressures that could have driven the increase of EQ in our pre-human ancestors. If one compares the brain of a spider monkey, for example, to that of a lemur, it's not simply a magnified version. Parts have been added and repositioned. The regions dedicated to smelling have shrunk, while the visual cortex has blown up. New maps have been added into the visual cortex, some offering the monkey new colors, others helping it to focus on important features in a moving object, like edges and shading. In a primitive brain the regions that perceive touch, vision, and sound, as well as the region that controls the body's movement, take up the entire cortex—the outer rind of the brain. In the evolution of primates, the brain looks like an inflating balloon with the painted-on regions separating from each other.

Primatologists have compared the EQs of monkeys in terms of their diet and found that the species that eat hard-to-get foods—termites, snails, seeds, nuts— or can eat a number of different foods have higher EQs than monkeys that settle for obvious grub or a limited selection of food. Perhaps there were great evolutionary rewards for keeping track of where different fruit trees grew, the time of year when they were ripe, and the special ways of handling each kind. That might explain the elaboration of the visual cortex. Another possibility arises when you consider the fact that the relative size of a monkey's neocortex is larger if it lives in larger societies. A marmoset living in a six-monkey band doesn't need a mind for faces, since each band member is pretty clear on its role. But in a large troop of monkeys like gelada baboons, where alliances are made and broken, where favors can be bestowed or sex refused, each individual has to keep an elaborate social map continually updated in its mind. Big societies can form among primates for a number of reasons. Food may be laid out across a landscape in such a way that monkeys can't avoid each other. Predators such as snakes and eagles may make it profitable to stick together in big groups, and in this association by necessity, each monkey will try to maximize its reproductive success through a sort of animal politics.

Every increase in EQ brought with it still more neural space that could be claimed by the neocortex, in which it could construct new systems for integrating information, for creating a richer representation of the world. By 5 million years ago, when the last common ancestor of human and chimps lived, our ancestors' brains could offer several signs of intelligence: a rudimentary sense of one's self, the ability to make complex tools (everything from termite skewers to leaf sandals), the ability to think of the world in abstract categories and to communicate with some simple gestures and calls. By 4 million years ago hominids proper were walking upright, and by 2.6 million years ago they were using stone tools to hunt and slice up meat. It was at about this time that brains began to explode far faster than ever before, an explosion that has continued into our own species.

Many researchers suspect that the beginnings of language—the expression of symbolic sounds or gestures, perhaps with some simple grammar—coincided with these two events. Hominids may have been forming a new kind of society, one made of huge bands in which the males sometimes hunted with tools and supplied females and children with meat, and language was the only thing that could keep these complex groups from falling apart. Whatever the reason, the other uses for language, such as allowing people to cooperate in ways no animals could or passing on their knowledge to their young, encouraged our brains to grow further. In a few hundred thousand years our brains had reached their current EQ, and along the way we found the means to give other humans a shard of our inner world.

Dorudon—the Everywhale of the Eocene—had a low EQ of about 0.42. Getting from *Dorudon* to a modern cetacean, such as a dolphin with an EQ over 4, is not easy. The road runs through the heavy shelling of the fossil-genes battles, which has made some stretches impassable. We can't say yet that baleen whales evolved independently from archaeocetes, or whether they are odontocetes in disguise. Making matters worse, barely any work has been done on the EQs of living mysticetes. For now it's possible to say meaningful things only about odontocetes. Few researchers have measured the brainpans of fossil whales, but what little evidence there is suggests that the odontocetes climbed up to their high EQs between about 25 and 15 million years ago. By 15 million years ago dolphins were as encephalized as they've ever been.

It wasn't echolocation per se that gave odontocetes their brains. Bats have a sonar system that in some ways is as sophisticated as a dolphin's, and they run it with a brain the weight of a raisin. In fact bat brains are actually small for a mammal of their size, and the echolocating species have brains half the size of ones who do without. Even that may not be the bare minimum, judging from the fact that a submarine can manage with a few thousand circuits. Rather than drive the evolution of the odontocete brain, echolocation probably simply laid the groundwork. Much evidence suggests that sperm whales and beaked whales were the first living odontocete lineages to branch off on their own. It's possible that the lives they lead now may still carry an echo of the earliest odontocetes. Early toothed whales were the first cetaceans that could hunt in sunless water, scanning their sonar beams for squid and fish a thousand feet below the surface of the oceans. The simplest way to pursue this kind of life would be to have a solitary existence, roaming alone and rising back up only for air. Yet the early odontocetes could not afford to live like bobbing sharks because as placental mammals they had to care for and nurse their young years, giving them the opportunity to learn the fine points of being a whale. They had no choice but to organize into pods.

Family units of sperm whales, made up mostly of adult females and their calves, make organized sweeps of the ocean on their dives (possibly sharing each other's echolocation), but stagger their swims so that there are always adults on the surface to take care of the calves. After a long session of dives, they will sometimes congregate at the surface, stroking one another's jaws and producing clicks, which, like a monkey's grooming or a human's small talk, may help keep the whales bonded together in their tight group. (Young males are booted out of the family after they get too old, and are left to wander among other female groups. They look for opportunities to mate, courting with a different set of click patterns.) This way of life works because sperm whales manage to live so long—perhaps sixty years—and so there are grandmothers that can rear the young of the family and carry on knowledge of the vast territories that sperm whales cover.

Complex social choreographies like these may have driven an early, apelike rise of EQ in odontocetes. Supporting this idea is the fact that among living toothed whales, higher EQs correspond to bigger pod sizes. But as the cetacean brain swelled, it did not take the primate path. In a monkey or ape, the regions of the brain that receive information from the senses are spread over the surface of the brain, separated by cortex without input from the outside, and the tissue is packed with a dense variety of neurons. In odontocetes the information from the senses all flows to parts of the cortex that are pushed together, while vast expanses of the brain's surface remain unclaimed. The cortex forms into five layers, three of which are practically a mystery to neurologists. The neurons that make them up are sparser, designed not so much for dense interconnections as for exquisite timing.

While diving for squid brings with it huge rewards for a whale, it may also put up a wall to any further increase in its brain size. Specializing in one plentiful animal, early whales may have mirrored the leaf-chewing langur monkeys, whose small brains may be determined by their simple diet. The deep diving that early cetaceans had to make to catch food may have been another constraint. When a whale plunges down thousands of feet into cold water, it needs to conserve oxygen and heat as carefully as possible. The brain has to be kept warm and demands a huge amount of oxygen, and so it's probably not worth it to such a whale to make it any larger. About 25 million years ago, however, a set of toothed whales apparently returned to shallow water, from which dolphins would emerge. In a flash (a flash much like our own) the ancestors of today's dolphins expanded their EQ to the levels they have today—by a factor of five in perhaps 5 to 10 million years.

Released from the constraints of deep dives, these cetaceans may have had to develop their brains because they were relatively small mammals with nowhere to hide. They had to stay in even larger, close pods in which individuals traded information constantly with clicks and sonar. In such large societies it paid to keep track of one's social life, to form coalitions more tangled than in earlier whale species. Another factor may have been the way these odontocetes changed their diet from deepwater squid to the many different species of fish that live in shallow seas, fish that they hunted cooperatively. With an increasingly complicated social life, an improved echolocation system, and a brain perfected for integrating information into maps, dolphins, like hominids, had inadvertently laid the groundwork for an abstracting mind. Relationships between things crystallized into a grammatical way of thinking of the world, and some odontocetes may have become able to share inner worlds by sharing sounds. Like our own mental powers, the intelligence of cetaceans is an exquisite product of correlated progression, in which not only raw anatomy plays a part, but society and thought as well, all carrying each other forward to a phenomenon our planet had never seen before.

No one should leave this story complacent in the triumph of human intelligence with its EQ of 7. Whales may turn out to be an exception to Jerison's rules because they don't need to mind their bodies as much. A fair part of their weight—as much as 30 percent—is blubber, which needs far fewer nerves than the same weight in muscle. Terrestrial mammals are forever struggling against gravity with careful measurements of balance and muscle contractions, but whales simply float, with up and down irrelevant to them except when they need to breathe. These two factors may mean that whales actually have far more brain space exempt from bodily groundskeeping than a conventional EQ would suggest. And if they should wiggle their way up through the ranks, we should also keep in mind the Training Game: we still hardly know what cetaceans can do with their brains. There was a time when some scientists were sure that in order to reach the fruits of intellect, vertebrates had to come on land. But underwater the mammalian brain seems to have found another orchard as well.

ON THE TRAILS OF MACROEVOLUTION

Back in 1841, when Richard Owen served as England's gate-keeper to all biology living and lithified, he stood before the British Association for the Advancement of Science and introduced his audience to the great vanished reptiles. "The period when the class of reptiles flourished under the widest modifications, in the greatest number, and the highest grade of organizations, is past," he told them, almost mournfully. He brought forward the animals he later named the dinosaurs, whose name meant "fearfully great lizards," great not only in their size but in their anatomy—the fine, complex workmanship in everything from their serrated teeth to their carefully cemented sacrums. And from the ocean he brought sea monsters, although not the disguised fossil whales of Albert Koch or frightening shipboard mirages. These were ancient fossils of enormous marine reptiles, some looking like yacht-sized swordfish, some with long necks the size of masts.

Owen had more on his mind than cataloging the fossil reptiles of Great Britain, though. He had been fighting the evolutionary ideas of Lamarck and Geoffroy for almost ten years as he studied animals ranging from the platypus to the lungfish and the chimpanzee. Now he took one further opportunity to stomp out a shoddy metamorphosis: one that could turn sea monsters into living crocodiles.

Sixteen years earlier, Geoffroy had studied the skull of an incongruous Mesozoic marine reptile called a teleosaur. It looked less like the sea monsters of its day than a fifteen-foot, thin-snouted crocodile that lived in the ocean. No living crocodile had its exact body plan, and yet as Geoffroy looked over their bones he wondered if teleosaurs and crocodiles might form part of a long evolutionary chain. The first link was the swordfish-shaped marine reptiles (called ichthyosaurs), which gave rise to the long-necked forms called plesiosaurs. From plesiosaurs in turn came the teleosaurs, from which came living crocodiles.

Owen plowed this idea under. He looked to the layers of rocks where teleosaurs, ichthyosaurs, and plesiosaurs had fossilized. If they had succeeded each other in the rocks as evolution demanded, he would admit that "some colour of probability might attach itself to this hypothesis." But the oldest fossils of the three extinct reptiles all appeared at the same time in the Mesozoic, and the ichthyosaurs and plesiosaurs actually outlived their supposed descendants, the teleosaurs.

It was Owen's tragedy to be right only up to a point. Fossils do indeed argue against Geoffroy's transformation, but the bones of marine reptiles that have been excavated since his day argue in favor of other transformations even more dramatic than he imagined. Teleosaurs were ancient marine crocodiles after all, descended with today's crocodiles from a common ancestor 220 million years old, a small slender crocodilian animal that could trot across dry land on legs shaped like a fawn's. Ichthyosaurs and plesiosaurs were not the ancestors of today's crocodiles but separate, extinct lineages descended from other terrestrial reptiles. In fact reptiles have moved back into the ocean sixteen times, and in most cases they've adopted a fishlike life—living permanently at sea, where they swam gloriously thanks to their long, tapered bodies and flipper-shaped limbs.

It was Darwin who recognized the natural patterns that explain the histories of marine reptiles and the rest of life, not Owen. And now, after 140 years of building on Darwin's original ideas and discovering crucial fossils, we can look at the transitions that marine reptiles took from land to sea, as well as those that many other lineages have made, and see a general form of macroevolution take shape. Between whales and the first tetrapod, for example, there are some remarkable coincidences. Think of their origins in grossly simplified form as an X. At the top left tip are mesonychidlike whale ancestors, which slide down their axis into the open ocean. At the bottom left tip are lobe-fins that make their way up the other axis onto land. The two animals that cross paths at the crux, *Acanthostega* and *Ambulocetus,* are suprisingly alike. They are both sudden ambushers hunting in shallow water, with long flat heads, snapping jaws, short limbs, and powerful tails. As whales moved farther into the ocean, they evolved fins and could use their tails to move. Their forelegs remained anchored to their shoulders, but their hind legs fell away, their hips going with them. In these ways the whale *Dorudon* is exactly like the lobe-fin *Eusthenopteron*.

But many of these reversals were illusions. Cetaceans made their flukes from connective tissue rather than fin rays, and rather than resorting to a fish's side-to-side oscillations they carried the mammal gait to water, becoming (along with sirenians) the only vertebrates in the ocean that swim by moving a tail up and down. Their ears reconciled with their jaws after 200 million years of separation, but rather than fusing into an old synapsid or fish form, they became part of a completely new way of hearing underwater. Far from being remade fishes, whales, like squid with their oversized eyes, are a vivid example of the quit point. They could not let their terrestrially evolved sense organs disappear, because they no longer had backup systems. Without a fish's lateral lines, they had to make their ears even more powerful. A fish with both lungs and gills could lose the ability to breathe air and not suffocate, but whales had to organize their lives around breathing, evolving countless little tricks of physiology and behavior to minimize their time at the water's surface. Whales are the frogs of the ocean, furiously coming up with graceful little inventions in order to live in a new home while held within tight evolutionary constraints.

The real similarity between whales and early tetrapods is not in their details but in the overall form of their metamorphosis. Each begins with a minor group of animals probing around the edges of a new ecosystem—be it the Devonian coastal wetlands or the shallow, salty Tethys Sea free of giant marine reptiles. These forerunners have inherited structures from distant ancestors that will have an unplanned and very different importance millions of years in the future: lungs for tetrapods, for instance, or long tails for mesonychids. As they tentatively explore this intermediate zone, they evolve new adaptations for their immediate conditions that will later turn out to be more exaptations, such as feet for moving underwater and perhaps cetacean ears that separated from the skull for hearing on land. New structures appear out of the blue, thanks to the rules by which all vertebrates develop. Genes accidentally duplicate, tissues come into contact with each other for the first time in embryos and bloom into something unexpected. Many of the same genes play a role in all of these innovations. Hox genes both patterned fingers and helped give whales a back that they could swim with.

At first the animals dabble, only partially transformed for their new environment. But it eventually becomes apparent that they have stumbled into empty terrain: for whales, an ocean robbed of its giant marine reptiles; for the first tetrapods, continents without a single vertebrate on board. They quickly branch into many forms, most of which are short-lived but a few of which manage to survive. Exaptations, now critical to their survival, look in hindsight like amazingly providential gifts. But although the animals have committed themselves so deeply to their new habitat that they can't survive in their former homes, much of macroevolution's work is still unfinished. Incrementally they add on new features, like a home owner who can only afford a shutter one year, a gutter the next. It takes dozens of millions of years to find a good strategy for harvesting food in their new habitat—be it the ability to eat a leaf, to engulf krill, or to see the ocean and all it contains with sound. Once they do, though, the animals spread, feast, and block off any other aspiring animals that would be like them. In 400 million years only a single lineage of vertebrates has come on land permanently. And although other predatory mammals like seals and otters spend much of their time at sea, only whales live in the ocean year-round.

A generous slice of this kind of change—from a fingerless lobe-fin to a tetrapod with sturdy limbs and no gills, or from a hoofed land mammal to a big fluked whale committed to the sea—took less than 15 million years in both cases. Compared to other stretches of evolutionary history, that's a fast clip. When Darwin traveled to the Galapagos Islands and saw iguanas leaping into the ocean to graze on seaweed, they had been leaping for perhaps as long as 15 million years with hardly any changes from the basic plan of terrestrial iguanas. In 1972 Stephen Jay Gould attacked the puzzle of varying evolutionary rates with one of his most controversial theories, which he proposed with Niles Eldredge of the American Museum of Natural History. In the fossil record, species often appear suddenly, hang on relatively unchanged for millions of years, and then vanish. Darwin had pointed out that when it comes to the past life of this planet, fossils are pebbles from a mountain, and he was sure that a full record would always show evolution obeying natural selection's gradual pace.

Gould and Eldredge suggested instead that the fossils could often be taken for their face value: new species often did branch suddenly away from older species, lingering for millions of years relatively unchanged until they became extinct—during which time newer species might abruptly branch away from them. Animals didn't go to sleep one night and in the morning find a new species running across their savanna. An isolated fragment of a population may be able to evolve in only fifty thousand years or so into a new species—too quickly for paleontologists to witness. If they find even a handful of fossils of a single species in so short a span of time they count themselves lucky. Chances would be overwhelming that those fossils would belong to the big, unchanging section of the population rather than from the small coterie that was actually evolving. If the new species thrived, it would eventually spread from its small birthplace and mingle with the ancestral species, and leave its own fossils which would seem to have appeared out of nowhere.

According to punctuated equilibrium (the name Gould and Eldredge put to their hypothesis), most changes happen as species originate, not during their lifetime. In other words, species are born from other species with certain traits which they carry to their extinction—just as an individual animal does. And just as the variation of individuals is the raw material that natural selection uses during microevolution, the variation from species to species may be the raw material for macroevolution. Species may compete, and they may give rise to new species at different rates. A lineage in which species don't speciate much might go extinct or linger as a living fossil, while others may be transformed—species by species—into unimaginable new forms.

When I talk to evolutionary biologists about punctuated equilibrium, I'm often surprised at the sting in their off-the-record remarks pro and con, twenty-five years after the theory was first hatched. In that time, some paleontologists have searched cliffs and mountainsides for unbroken sketches of fossil-rich rocks where they can test this idea. In many cases new lineages do seem to branch suddenly from one species to another, while in some others they drift apart more gently. Meanwhile, some researchers who have been trying to measure evolution's natural pace in living animals have been surprised at how quickly it can move. Evolution can change an animal's body rapidly, as Gould and Eldredge argued, but the change doesn't have to happen in conjunction with the origin of a new species.

The sting comes from the fact that testing punctuated equilibrium is an unfinished business. Yet no matter how it survives, it has already had one clear effect: it prodded paleontologists to invent new ways to test the patterns of macroevolution. It has become apparent to all sides of the debate, for example, that many of the long evolutionary coasts suggested by the fossil record are real, and deserve an explanation. To some gradual-minded scientists, a record of fossils with a 100,000-year resolution that looks like stasis may actually be hiding a riot of generation-by-generation change that ends up not going very far in any one direction. To others stasis means that an animal's surroundings simply make no demands on it to change for a long time. Those who prefer a punctuated view point out that the climate—probably the most important part of an animal's surroundings—can dramatically swing many times over the lifetime of a species, and yet the species will often seem unaffected. Drastic change is rare. In the face of a stampeding glacier, it's easier for a species of beetle to head south than stay and adapt to the new climate.

With the handful of primitive tetrapods and early whales known today, it's impossible to say whether any of the shifts from one of these species to another happened gradually or suddenly. In a sense, the answer doesn't matter, because punctuated equilibrium was not designed to explain macroevolution at the scale of these sorts of transitions. In one carefully studied example of punctuated equilibrium, Alan Cheetham at the Smithsonian Institution examined bryozoans, tiny animals that form matlike colonies underwater. Over the last 20 million years, species of the Caribbean genus *Metrarabdotos* have sprouted off with a reliable suddenness into new forms. But while a new species may have had slightly larger brood chambers or smaller feeding orifices, it remained very much a *Metrarabdotos* bryozoan.

Lobe-fins took a much longer anatomical journey, and not in one single leap, as evidenced by all the fossils of tetrapodlike fish and fishlike tetrapods that paleontologists have found. Regardless of whether particular changes along the way happened within a single species or in punctuations, they combined into an evolutionary fugue. For a time macroevolution concerned itself mainly with reshaping the tetrapod skull roof while the limbs changed little, but later the braincase and limbs began altering quickly. The many changes necessary to build a tetrapod accrued over millions of years, and not in a single burst.

More and more sets of transitional fossils are emerging, and this pattern among tetrapods holds true among them. Along the coasts of the Cretaceous oceans swam gigantic marine reptiles known as mosasaurs. Reaching forty-five feet long, these cold-blooded leviathans swam by undulating a whiplike tail and maneuvering with paddle-shaped feet. A preliminary look at their fossils suggests that the mosasaurs suddenly evolved from some reptile on land and stayed fixed in their new incarnation until they became extinct 65 million years ago. But in 1993, when Michael DeBraga and Robert Carroll at McGill University in Montreal looked over all the fossil evidence that has built up over the years, they were able to show that this is not the real history.

Mosasaurs descend from the same group of lizards that includes the Komodo dragon and other monitor lizards—and their ancestors probably diverged from this stock 155 million years ago. Over the course of 65 million years they changed into amphibious, crocodilelike animals. They could still walk on land, but they were large, with shrunken hips and shoulders, a jaw with a new hinge that let them catch fish more effectively, and a long tail for swimming. After this leisurely pace the ancestors of mosasaurs went through a 3-million-year burst of evolution that produced the first true mosasaurs—most notably by making their arm bones small and rigid, while stretching out their fingers. In another 3 million years three families of mosasaurs emerged, which then existed for 20 million years until their extinction.

DeBraga and Carroll counted up the traits that arose in mosasaurs as a crude evolutionary clock. They found that evolution moved slowly among the ancestors of mosasaurs, sped up drastically during their origin and their split into three major lineages, and slowed down again afterward, although not as slowly as among the pre-mosasaurs. The actual origin of mosasaurs was not the time at which evolution was working the fastest. After the first mosasaur had appeared and the three major lineages were beginning to diverge, there was twice as much evolution as during the time when the first mosasaur evolved. As with whales, a lot of the most important adaptations to water came only long after mosasaurs had become committed to the sea—their limbs becoming more paddlelike, their tails longer, their jaws more mobile—and happened independently in the three lineages.

The spurts at which these transitions take place might be the result of some kind of steady selection pressure applied on the members of a given species, or it might be the result of the patterns in which new species branch off and eventually go extinct. Whales did well enough amphibiously, but the rewards of being utterly oceangoing were enormous. But those rewards probably existed only because giant marine reptiles had been plucked out of the oceans by a catastrophe. Some researchers argue that macroevolution is only microevolution left running for a few thousand generations. But without understanding the external factors that set the obstacle course for natural selection to run, we can't understand the history of life. While species can withstand much of the change in their environment, they do fall victim in huge numbers to violent shifts. Oceans suddenly robbed of oxygen or overwhelmed by carbon dioxide, abrupt atmospheric greenhouses and icehouses can all knock out common, dominant organisms, leaving ecosystems open for new experiments by the survivors. Modern whales for example may have gotten their own start when a rapid change in climate drove archaeocetes extinct.

In addition to external factors, however, natural selection also has to obey internal ones, since it can only select from what's offered. A species' own particular mesh of genes and rules of development decide what kind of variation will turn up in each generation. A foot isn't necessarily the ideal structure for walking, created from the void; it comes from a twist in the pattern of Hox genes—genes that, like ancient dictators, have dominated animal evolution for a billion years, back when jellyfishlike creatures had the run of the oceans. Such rules of development may resist changes and may therefore be responsible for the stasis that lineages of life often experience. You can think of the genes that shape an embryo as creating an abstract sort of canal. As an embryo grows, as its tissues take form, it moves along the canal down toward the normal form for its species. Abuses to the embryo—poisons that get into an egg, or mutations to genes—push it up the sides of the canal, toward a different kind of anatomy. But most pushes are too weak, and the embryos fall back into the canal and are born as ordinary creatures. While Neil Shubin found a lot of variation in the hands of his frozen salamanders, the mutations never derailed the program of genes that builds their hands. Whales still have five-fingered hands, perhaps because the Hox genes that pattern them help shape so many other parts of the body. The canal is simply too deep.

Sometimes thinking about what macroevolution never created is just as useful as considering what it has. The giant Mesozoic marine reptiles worked their way into many ecological niches: some attacked the slow-moving, hard-shelled nautilus and its relatives, others chased fish, and still others were big enough to attack other aquatic reptiles. But as Rachel Collins of the University of Washington and Christine Janis of Brown University have pointed out, in the 180 million years that these sea monsters lived, macroevolution apparently never turned any of them into filter feeders like baleen whales. External factors don't seem to have been at play in this case, because the rocks show signs that during several spans of the Mesozoic the oceans supported rich supplies of plankton. Engulfing clouds of plankton or schools of small fish would have been an excellent way of life, and yet these animals never achieved it. Here we have a case where we can't simply say that natural selection forbade baleenlike reptiles.

The full history of lineages helps decide who will evolve into what. Whales were able to become filter feeders thanks to evolutionary episodes that took place over 200 million years before the first cetaceans were born. As the synapsid ancestors of mammals went through a correlated progression of changes in how they ate, breathed, ran, and reproduced, they separated the paths that air and food took. The nasal passages of synapsids grew to the back of their mouths, and at the same time, the tongue and soft palate evolved so that when they chewed, a seal kept food from getting caught in the airway. Without the seal in the back of its mouth, the synapsid would choke.

These seals allowed some descendants of the synapsids—the baleen whales—to take into their mouths enormous amounts of water that they could ram out with their tongues. If marine reptiles ever tried to open their mouths and take in a few hundred gallons of water, they would have no way to keep it from pouring into their guts or lungs. Aside from the problems a flooded stomach would cause, the reptile couldn't use its tongue to ram it back out of its mouth. And once reptiles were in a position where mouth seals would be useful, they couldn't evolve them. These structures didn't evolve in isolation in synapsids but as part of the dense web of changes that occurred during their evolution on land. Constrained by their history, marine reptiles never became filter feeders, and that may be the reason that it was whales that became the biggest animals ever to live.

There is always a little sadness mixed into great discoveries because they take away some of the confusion that brightens life. Before *Acanthostega*, before *Ambulocetus*, and before all the other clues that have coalesced in the past few years, a metamorphosis from a fish into a tetrapod or a mesonychid into a whale had the feel of one of Ovid's magical transformations. Now these metamorphoses join ordinary reality. For consolation consider that with the study of animals such as whales and early tetrapods, we have for the first time a model for how macroevolution carries life through the most extraordinary transitions, a model which can be applied to other transitions scattered across the tree of life.

The tree we now see is quite different from the ones that the early Darwinians drew. Ernst Haeckel published one with single-celled organisms at the base of its trunk and leafy boughs branching off at what he deemed higher and higher levels of organization, rising from invertebrates to vertebrates to mammals, until it reached Man perched at the top like a Christmas tree angel. The new phylogeny of life—built on anatomy, genes, and cladistics—looks like a computer-generated thicket. There are so many lineages now recognized that even vertebrates as a group get lost in the thorns. Pull back a little from our own region of the thicket and you can see our more distant relatives, and the more ancient common ancestors that gave rise to us all. The echinoderms—starfishes and sea urchins—come into view. Pull back farther still and another major branch of animals—which includes arthropods and crustaceans—emerges.

Along the nodes of these branches you can read thousands of transitions just as you can among the early tetrapods or whales, such as how insects came to land and how they came to fly. The story of insect evolution, which still waits to be fully fleshed out, will certainly be unique—you only need to consider a fly's exoskeleton and compound eyes, never mind the rest of its bizarre anatomy. But it now looks as if the insects' story may end up having a remarkable resemblance to our own: despite our radically different bodies, we appear to construct our eyes, limbs, and trunks with almost identical networks of genes.

Pull back further still and other animals come into view—the corals, the jellyfish, the many unaccountable animals looking like fronds and hockey pucks and spindles that came and went between 600 and 500 million years ago. At this point the branches that represent humans, sharks, and anacondas are looking like the split ends of a hair. Now you can see where the animal kingdom's branch first diverged from those of organisms like fungi over a billion years ago. Here are the plants coming into view as well, organisms that can often choose between cloning and sex for reproduction, with bark instead of bone—and therefore carrying their own evolutionary histories. Draw back more and the first multicellular organisms appear, which made the transition from solitary cell to collective. And finally microbes—the archea and bacteria—now loom, with branches so long and dense that the entire animal kingdom collapses down to a sprout. Here among the microbes genes get traded as casually as business cards among salesmen, and parasites plunging into hosts are turned into integral organs of the host itself. In these regions macroevolution doesn't invent legs or warm-bloodedness. It engineers proteins that let bacteria survive in a block of salt or hold together in boiling water. Even in these parts of life's thicket you can see passages that echo our own: a lineage of microbes that lives in the open ocean around Antarctica is nestled in among ones that live only in a hot spring in Yellowstone or in the topsoil of Indiana; another invented photosynthesis to harness sunlight, only to have it stolen later by algae.

If we had learned of these kinds of transitions a century ago they might have seemed too remote and radical to ever be understood. The story of each of these transformations hides its own unexpected details, as startling as the skyward eyes that sat on top of our ancestors' heads or the delicate toes that turned up in the equation of a whale. Yet if the transformations we do understand serve as any guide, some common themes will emerge. Rather than simplistic leaps of evolution-on-demand, they will be extensions of histories reaching back hundreds of millions of years. Many will prove to be based on revolutions in the way organisms grow, which then set up new ground rules for natural selection. They will often turn out to depend on the symphony of correlated progression, and on the gradual accretion of adaptations that slowly ooze into new, seemingly providential functions. Their stories will unfold in the perversely twisted arena of real-world ecosystems, where eyes swell before vanishing, where extinctions can suddenly switch on a Vacancy sign for new tenants. And in years to come more and more of their stories will be told, because biologists are learning the crafts—the sifting of branches, the reading of genes and fossils, and all the rest—that will make them into nature's new Ovids.

EVOLUTIONARY CHRONOLOGY

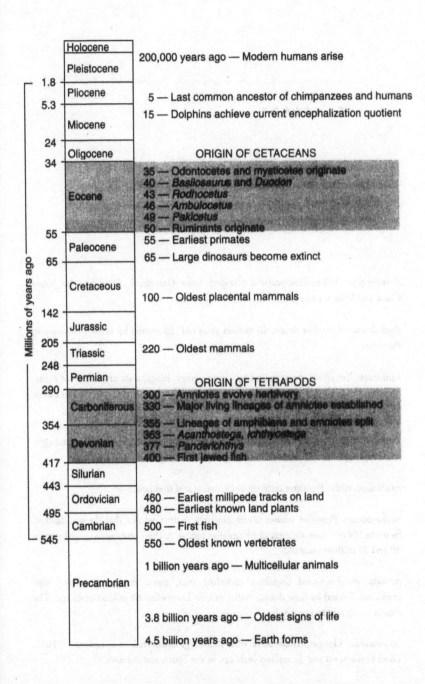

Holocene
200,000 years ago — Modern humans arise

Pleistocene

1.8
Pliocene 5 — Last common ancestor of chimpanzees and humans
5.3
15 — Dolphins achieve current encephalization quotient
Miocene

24
Oligocene ORIGIN OF CETACEANS
34
35 — Odontocetes and mysticetes originate
40 — *Basilosaurus* and *D-odon*
43 — *Rodhocetus*
Eocene 46 — *Ambulocetus*
49 — *Pakicetus*
50 — Ruminants originate
55
55 — Earliest primates
Paleocene
65 65 — Large dinosaurs become extinct

Cretaceous
100 — Oldest placental mammals

142
Jurassic

205
Triassic 220 — Oldest mammals

248
Permian ORIGIN OF TETRAPODS
290
300 — Amniotes evolve herbivory
Carboniferous 330 — Major living lineages of amniotes established
354
355 — Lineages of amphibians and amniotes split
363 — *Acanthostega, Ichthyostega*
Devonian 377 — *Panderichthys*
400 — First jawed fish
417
Silurian

443
Ordovician 460 — Earliest millipede tracks on land
480 — Earliest known land plants
495
Cambrian 500 — First fish
545
550 — Oldest known vertebrates

1 billion years ago — Multicellular animals

Precambrian

3.8 billion years ago — Oldest signs of life

4.5 billion years ago — Earth forms

Millions of years ago

Acanthostega: 363-million-year-old tetrapod from Greenland. Described by Jenny Clack and Mike Coates.

Ambulocetus: Otterlike whale, 46 million years old, discovered by Hans Thewissen in Pakistan.

Amniotes: Tetrapods characterized by, among other things, a set of embryonic membranes inside either an egg or the uterus. Includes all living terrestrial vertebrates except amphibians.

Amphibians: One of the two major living branches of tetrapods. Lay jelly-coated eggs. Includes frogs, salamanders, and caecilians.

Anthracobunids: Primitive relatives of elephants and sirenians.

Archaeocetes: Primitive whales (more precisely, extinct whales that do not descend from the last common ancestor of whales alive today). Archaeocete fossils range between 49 and 37 million years old.

Artiodactyls: Even-toed ungulates including pigs, cows, deer, camels, and hippopotami. United by their double-pulley ankles. Diversified 65 million years ago. The closest living relatives to whales.

Basilosaurus: Elongate archaeocete with small legs discovered in Louisiana in 1832. Lived between 40 and 37 million years ago in the Tethys and Atlantic.

Caecilians: Legless amphibians that lay their eggs underground. Some burrow like earthworms, others live in leaf litter, and still others are aquatic.

Cetaceans: Whales, dolphins, and porpoises, all of which descended from a terrestrial mammalian common ancestor about 50 million years ago.

Cladistics: A method of organizing life on an evolutionary tree (a cladogram) by comparing traits shared by different sets of species.

Coelacanths: Lobe-finned fishes that live in the deep ocean off the coast of Africa. The sole survivors of a 400-million-year-old lineage.

Cuvier, Georges: Founder of paleontology and the leading comparative anatomist of the early nineteenth century (1769–1832). Opponent of evolution, believed that biological function determined form.

Dorudon: Fifteen-foot-long archaeocete that lived 40 to 37 million years ago. Close to the ancestry of living whales.

Echolocation: The ability to perceive one's surroundings from the echoes of one's calls. Perfected by odontocetes and bats.

Encephalization quotient (EQ): The ratio of an animal's actual brain size to the expected size for a typical animal of this same body weight. A useful index for comparing the intelligences of different species.

Eusthenopteron: Devonian lobe-fin, closely related to tetrapods.

Exaptation: A trait that finds a new function during the course of evolution. For example, limbs and digits evolved for locomotion underwater; they became an exaption for walking on land. Sometimes called (misleadingly) preadaptation.

Gene: Segment of DNA that constructs a protein.

Homeotic/Hox genes: Genes that help guide construction of the axis and limbs of an animal. Hox genes are the homeotic genes of vertebrates.

Hynerpeton: 367-million-year-old tetrapod from Pennsylvania discovered by Ted Daeschler. Oldest tetrapod that definitely lacked gills as adults.

Ichthyostega: 363-million-year-old tetrapod discovered by Gunnar Säve-Söderbergh.

Lamarck, Jean-Baptiste: French naturalist (1744–1829). Proposed an early theory of evolution based on inheritance of acquired characters and internal striving.

Lepidosiren: Lungfish first discovered in Brazil.

Linnaeus, Carolus: Swedish naturalist (1707–1778) who established the standard classification system for life (species, genus, family, etc.).

Lobe-finned fish: One of the major lineages of fish. Characterized by fins with a large axis of skeletal bone. Includes coelacanths, lungfish, *Eusthenopteron,* and tetrapods.

Melon: Fat deposit in the forehead of odontocetes that focuses sound waves for echolocation.

Mesenchyme cell: A primordial type of cell in the limb bud of a tetrapod embryo which condenses into cartilage, which in turn is replaced by bone.

Mesonychids: Meat-eating ungulates that lived from 65 to 34 million years ago. Closest relatives of whales.

Morphogenetic field: A region of an embryo in which uniformly distributed chemical reactions or mechanical forces can create complex patterns.

Mysticetes: Baleen whales (includes humpback whales, blue whales, right whales, gray whales, and fin whales).

Nucleotide: One of four kinds of information-bearing molecules in DNA.

Odontocetes: Toothed whales (includes dolphins, beluga whales, killer whales, sperm whales, narwhals, and beaked whales).

Ontogeny: The course of an embryo's development.

Owen, Richard: British anatomist (1804–1892) who proposed the vertebrate archetype and became a public opponent of Darwinism.

Pakicetus: Oldest and most primitive whale known, at 49 million years of age. Discovered by Philip Gingerich in 1979.

Perissodactyls: Odd-toed ungulates, including horses, tapirs, and rhinoceroses.

Phylogeny: The evolutionary history of a species.

Proteins: Long-chained molecules made of amino acids. Assembly is determined by genes.

Rodhocetus: 43-million-year-old archaeocete. Oldest known tail-driven whale.

Ruminants: Artiodactyls with a rumen housing colonies of bacteria, and a stomach with special lysozyme proteins that allow them to digest plant material efficiently. Includes cows, sheep, and goats.

Sirenians: Marine mammals such as dugongs and manatees.

Stapes: Large bone in the middle ear. Formerly the hyomandibular in lobe-finned fish.

Synapsids: The group of amniotes that includes mammals.

Tetrapods: Terrestrial vertebrates (including secondarily aquatic groups such as whales), all of which descended from a lobe-fin ancestor about 370 million years ago.

Tulerpeton: Six-fingered, 355-million-year-old tetrapod from Russia. Possibly the closest relative of amniotes.

Ungulates: Hoofed mammals. Includes artiodactyls, perissodactyls, mesonychids, and elephants.

NOTES

INTRODUCTION: LIFE'S WARPS

3   If I were foolish enough . . . J. B. Wyngaarden, L. H. Smith, and J. C. Bennett, eds., *Cecil Textbook of Medicine* (Philadelphia: W. B. Saunders, 1979).

3   The yellowtail's gills . . . G. M. Hughes, "General anatomy of the gills," *Fish Physiology* 1984, no. 10A: 1–72.

4   To understand this tortuous layout . . . W. Gilbert, *Developmental Biology*, 4th ed. (Sunderland, Mass.: Sinauer, 1994).

6   When Aristotle walked . . . Aristotle, *Historia Animalium*, trans. A. L. Peck (Cambridge: Harvard University Press, 1965); quote from p. 589.

6   The properties of air and water . . . P. Dejours, L. Bolis, C. R. Taylor, and E. R. Weibel, eds., *Comparative Physiology: Life in Water and on Land* (Padova: Liviana Press, 1987).

7   Microevolution has been so well studied . . . For a popular examination of microevolution, see J. Wiener, *The Beak of the Finch: A Story of Evolution in Our Time* (New York: Knopf, 1994). For examples of microevolution as an applied science, see P. H. Harvey et al., eds., *New Uses for New Phylogenies* (Oxford: Oxford University Press, 1996).

CHAPTER 1. AFTER A LOST BALLOON

Interviews with Hans Bjerring, Jenny Clack, Michael Coates, Kevin Padian, Martin Rudwick, Phillip Sloan, and Keith Thomson.

9   In a basement laboratory . . . This account of the life and work of Sir Richard Owen is based on A. Desmond, *Archetypes and Ancestors: Paleontology in Victorian London, 1850–1875* (Chicago: University of Chicago Press, 1984); Desmond, *The Politics of Evolution: Morphology, Medicine, and Reform in Radical London* (Chicago: University of Chicago Press, 1989); R. M. Macleod, "Evolutionism and Richard Owen, 1830–1868: An episode in Darwin's century," *Isis* 1965, 56:259–280; R. Owen, *The Life of Richard Owen* (New York: D. Appleton, 1894); K. Padian, "The rehabilitation of Sir Richard Owen," *Bioscience* 1997, 47:446–453; N. A. Rupke, *Richard Owen: Victorian Naturalist* (New Haven: Yale University Press, 1994); and P. R. Sloan, "On the edge of evolution," introduction to R. A. Owen, *The Hunterian Lectures in Comparative Anatomy, May and June 1837* (Chicago: University of Chicago Press, 1992).

11   In 1830 a sixty-one-year-old French baron . . . Details of Cuvier's life can be found in W. Coleman, *Georges Cuvier, Zoologist* (Cambridge: Harvard University Press, 1964); and D. Outram, *Georges Cuvier: Vocation, Science, and Authority in Post-revolutionary France* (Manchester: Manchester University Press, 1984).

12   "Come play . . ." Quoted in T. A. Appel, *The Cuvier-Geoffroy Debate: French Biology in the Decades before Darwin* (Oxford: Oxford University Press, 1987) p. 31.

12   People had collected fossils . . . M. J. S. Rudwick, *The Meaning of Fossils: Episodes in the History of Palaeontology* (New York: Science History Publications, 1976).

13   "Is not Cuvier . . ." Quoted in Appel, *The Cuvier-Geoffroy Debate*, p. 46.

14   Geoffroy and Cuvier had drifted . . . See Appel, *The Cuvier-Geoffroy Debate*, for a full account.

15   "Unity of Plan . . ." quoted from Appel, *The Cuvier-Geoffroy Debate*, p. 226.

16   While Owen was busy . . . The early scientific history of the lungfish is outlined in D. E. Rosen et al., "Lungfishes, tetrapods, paleontology, and plesiomorphy," *Bulletin of the American Museum of Natural History* 1981, 167:159–276; and E. B. Conant, "An historical overview of the literature of Dipnoi: Introduction to the bibliography of lungfishes," *Journal of Morphology Supplement* 1986, 1:5–13.

16   "victims . . . of Natterer's too passionately executed chase . . ." Quoted from Conant, "An historical overview," p. 7.

17   In June 1837 . . . R. A. Owen, "On a new species of the genus *Lepidosiren* of Fitzinger and Natterer (*L. annectens*)," *Proceedings of the Linnean Society of London* 1839, 1:27–32; and "Description of the *Lepidosiren annectens*," *Transactions of the Linnean Society of London* 1841, 18:327–362.

17   Darwin had gone to the University of Edinburgh . . . A. Desmond and J. Moore, *Darwin: The Life of a Tormented Evolutionist* (New York: W. W. Norton, 1991).

18   The African creature he wanted to call *Protopterus* . . . Actually, Owen would ultimately have a little satisfaction on this count. The lungfish of Brazil and Africa in fact belong to different genera. The African genus contains five species, which all look essentially like the one that Owen studied. While they each have their own species name, they all belong to the genus *Protopterus*.

18   "Since the time of *Ornithorhynchus* . . ." Owen, "On a new species," p. 27.

19   "If indeed, the species had been known only by its skeleton . . ." Owen, "Description of *Lepidosiren*," p. 350.

19   "In the organ of smell . . ." Owen, "Description of *Lepidosiren*," p. 362.

20   a deeper relationship, which Owen called homology . . . For an introduction, see A. L. Panchen, "Richard Owen and the concept of homology," in *Homology: The Hierarchical Basis of Comparative Biology*, ed. B. K. Hall, pp. 21–62 (New York: Academic Press, 1994).

20   "the same organ . . ." Quoted in Rupke, *Richard Owen*, p. 134.

21   "But from this Epicurean slough . . ." R. A. Owen, *On the Origin of Limbs*, p. 40 (London: Van voorst, 1849).

22   In 1859 . . . Among the flood of recent considerations of Darwin's work, see Desmond and Moore, *Darwin*; E. Mayr, *The Growth of Biological Thought* (Cambridge: Harvard University Press, 1982); and D. Ospovat, *The Development of Darwin's Theory* (Cambridge: Cambridge University Press, 1981).

22   "I look at Owen's Archetypes . . ." Quoted from Desmond and Moore, *Darwin*, p. 368.

23   "guess-endeavours . . ." Quoted from Rupke, *Richard Owen*, p. 245.

23   "I used to be ashamed . . ." Quoted from Desmond and Moore, *Darwin*, p. 596.

23   "*Our* ancestor . . ." Quoted from Desmond and Moore, *Darwin*, p. 505.

26   This combination of Darwin's ideas . . . For a summary of neodarwinism, see, for example, M. Ridley, *Evolution* (Boston: Blackwell, 1993).

27   As more specimens of *Lepidosiren* were studied . . . J. Hogg, "On the existence of branchiae in the young Caecilae; and on a modification and extension of the branchial classification of the Amphibia," *Annals and Magazine of Natural History* 1841, 7:355–363; W. Jardine, "Remarks and on the structure and habits of *Lepidosiren annectens*," *Annals and Magazine of Natural History* 1841, 7:21–26; A. G. Melville, "On the *Lepidosiren*," *Report of the British Association for the Advancement of Science* 1847, 17:78.

27   Later scientists like the Irish anatomist Robert M'Donnel . . . R. M'Donnel, "Observations on the habits and anatomy of the *Lepidosiren annectens*," *Natural History Review* 1860, 7:93–112. Quotes from pp. 95, 112.

27   A third continent . . . G. Krefft, "Description of a gigantic amphibian allied to the genus *Lepidosiren*, from the Wide-Bay District, Queensland," *Proceedings of the Zooligical Society of London* 1870, 221–224.

28   As one skeptic wrote . . . A. Gunther, "Description of *Ceratodus*, a genus of Ganoid fishes recently discovered in rivers of Queensland, Australia," *Philosophical Transactions of the Royal Society of London* 1871, 161:511–571.

29   For closer kin . . . P. Bowler, *Life's Splendid Drama: Evolutionary Biology and the Reconstruction of Life's Ancestry* (Chicago: University of Chicago Press, 1996); E. D. Cope, "On the phylogeny of the vertebrata," *Proceedings of the American Philosophical Society* 1892, 30:278–285; W. K. Gregory, "The limbs of *Eryops* and the origin of paired limbs from fins," *Science* 1911, 33:508–509; T. H. Huxley, "On the application of the laws of evolution to the arrangement of the Vertebrata and more particularly on the Mammalia," *Proceedings of the Zoological Society of London* 1880, 649–662; and D. M. S. Watson, "The evolution and origin of amphibians," *Philosophical Transactions of the Royal Society of London* 1926, B 114:189–257.

30   "Between the oldest known Amphibia . . ." Quoted from W. K. Gregory, "Present status of the problem of the origin of the Tetrapoda with special reference to the skull and paired limbs," *Annals of the New York Academy of Sciences* 1915, 26:317–383.

30   A doomed balloon ride . . . Andrée's ride is described in P. Berton, *The Arctic Grail* (New York: Penguin, 1988).

31   His fellow Swedish scientists searched for him . . . E. Jarvik, "The Devonian tetrapod *Ichthyostega*," *Fossils and Strata* 1996, 40:1–213.

32 "scales of a fish-like vertebrate . . ." Jarvik, "The Devonian tetrapod," p. 5.

32 Back in Stockholm . . . G. Säve-Söderbergh, "Preliminary note on Devonian ste-gocephalians from East Greenland," *Meddelesler om Grønland* 1932, 94(7): 1–105.

33 He could see that its legs . . . E. Jarvik, "The oldest tetrapods and their forerun-ners," *Scientific Monthly* 1955, 80:141–154; and *Basic Structure and Evolution of Vertebrates* (London: Academic Press, 1980).

CHAPTER 2. LIMITLESS AIR, HO!
Interviews with Per Ahlberg, Jenny Clack, Michael Coates, Walter Cressler, Ted Daeschler, and Keith Thomson.

35 "but fishes . . ." R. S. Lull in J. Barrell et al., *The Evolution of the Earth and Its In-habitants* (New Haven: Yale University Press, 1920) p. 119.

36 According to a geologist named Joseph Barrell . . . The two key works of Barrell are "Dominantly fluviatile origin under seasonal rainfall of the Old Red Sand-stones," *Bulletin of the Geological Society of America* 1916, 27:345–386, and "In-fluence of Silurian-Devonian climates on the rise of air-breathing vertebrates," *Bulletin of the Geological Society of America* 1916, 27:387–436.

36 "Was it in the physical condition . . ." Barrell, "Dominantly fluvatile," p. 346.

37 "To breathe air . . ." Barrell, "Influence," p. 409.

37 Romer added some biological realism . . . Romer laid out his hypothesis in sev-eral papers and retold it in books and addresses. See for example his *Vertebrate Pa-leontology* (Chicago: University of Chicago Press, 1933); "The early evolution of land vertebrates," *Proceedings of the American Philosophical Society* 1956, 100:157–167; "Tetrapod limbs and early tetrapod life," *Evolution* 1958, 12:365–369; and "Major steps in vertebrate evolution," *Science* 1967, 158:1629–1637.

37 Walter Bock . . . See W. J. Bock, "Preadaptation and multiple evolutionary path-ways," *Evolution* 1959, 13:194–211; and "The role of adaptive mechanisms in the origin of higher levels of organization," *Systematic Zoology* 1965, 14:272–287.

38 Stephen Jay Gould of Harvard and Elizabeth Vrba . . . S. J. Gould and E. Vrba, "Exaptation—a missing term in the science of form," *Paleobiology* 1982, 8:4–15.

39 At the time, Europe was pinning . . . D. L. Woodrow, ed., *The Catskill Delta* (Boulder, Colo.: Geological Society of America, 1985).

42   When Daeschler published his new tetrapod . . . See E. B. Daeschler et al., "A Devonian tetrapod from North America," *Science* 1994, 265:639–642.

44   Red rock can be formed . . . See P. D. Krynine, "The origin of red beds," *Transactions of the New York Academy of Sciences* 1949, 11:60–68. See also R. F. Inger, "Ecological aspects of the origins of the tetrapods," *Evolution* 1957, 11:373–376; and G. L. Orton, "Original adaptive significance of the tetrapod limb," *Science* 1954, 1042–1043.

44   The other part of the old scenario . . . Among Keith Thomson's key papers are "The biology of the lobe-finned fishes," *Biological Reviews of the Cambridge Philosophical Society* 1969, 44:91–154; "The environment and distribution of Paleozoic sarcopterygian fishes," *American Journal of Science* 1969, 267:457–464; and "The ecology of Devonian lobe-finned fishes," in *The Terrestrial Environment and the Origin of Land Vertebrates*, ed. A. L. Panchen, pp. 187–222 (London: Academic Press, 1980).

45   Before about 480 million years ago . . . W. N. Steward and G. W. Rothwell, *Paleobotany and the Evolution of Plants*, 2nd ed. (Cambridge: Cambridge University Press, 1993); G. J. Retallack, "Early forest soils and their role in Devonian global change," *Science* 1997, 276:583–585; and P. Kenrick and P. Crane, "The origin and early evolution of plants on land," *Nature* 1997, 389:33–39.

48   an interesting jaw from Australia . . . K. S. W. Campbell and M. W. Bell, "A primitive amphibian from the Late Devonian of New South Wales," *Alcheringia* 1977, 1:369–381.

49   There was a tetrapod by the name of *Pholiderpeton* . . . J. A. Clack, "*Pholiderpeton scutigerum* Huxley, an amphibian from the Yorkshire coal measures," *Philosophical Transactions of the Royal Society of London* 1987, B 318:1–107.

50   What to do? . . . Clack's discovery of *Acanthostega* is detailed in early papers such as J. A. Clack, "New material of the early tetrapod *Acanthostega* from the Upper Devonian of East Greenland," *Paleontology* 1988, 31:699–724; as well as M. I. Coates, "The Devonian tetrapod *Acanthostega gunnari* Jarvik: Postcranial anatomy, basal tetrapod interrelationships, and patterns of skeletal evolution," *Transactions of the Royal Society of Edinburgh: Earth Sciences* 1996, 87:363–421; and C. Zimmer, "Coming onto the land," *Discover*, June 1995.

CHAPTER 3. HOW TO MAKE A HAND

Interviews with Pere Alberch, Jenny Clack, Michael Coates, David Irwin, Neil Shubin, Philip Sloan, and Guenter Wagner.

58   "one pair of limbs . . ." Quoted from R. A. Owen, *On the Origin of Limbs*, p. 9.

59   Darwin himself had only a hunch . . . E. Mayr, *The Growth of Biological Thought.*

60   The sea squirt . . . H. Gee, *Before the Backbone* (New York: Chapman & Hall, 1996).

60   In 1866 it showed up . . . For the rise and fall of recapitulationism, see Desmond and Moore, *Darwin;* S. J. Gould, *Ontogeny and Phylogeny* (Cambridge: Harvard University Press, 1977); C. Patterson, "How does phylogeny differ from ontogeny?" in *Development and Biology,* ed. B. C. Goodwin, N. Holder, and C. C. Wylie, pp. 1–31 (Cambridge: Cambridge University Press, 1983); and M. K. Richardson, "Heterochrony and the phylotypic period," *Developmental Biology* 1995, 172:412–421.

61   Fins come in many shapes . . . N. Shubin, "The evolution of paired fins and the origin of tetrapod limbs," *Evolutionary Biology* 1995, 28:39–86.

61   A British paleontologist named D. M. S. Watson compared *Eusthenopteron* . . . D. M. S. Watson, "The evolution and origin of amphibians," *Philosophical Transactions of the Royal Society of London* 1926, B 114:189–257.

62   By some estimates, over 70 percent of evolutionary transformations . . . P. Alberch, "'Ontogeny and Phylogeny' revisited: 18 years of heterochrony and developmental constraints," in *Biodiversity and Evolution,* ed. R. Aria et al., pp. 229–249 (Tokyo: National Science Museum, 1995).

63   Some salamanders, for instance . . . R. A. Raff and G. A. Wray, "Heterochrony: Developmental mechanisms and evolutionary results," *Journal of Evolutionary Biology* 1989, 2:409–434.

63   Turing . . . is best known . . . A. Hodges, *Alan Turing: The Enigma* (New York: Simon & Schuster, 1983).

64   In the last few years of his life . . . A. Turing, "A chemical model of morphogenesis," *Philosophical Transactions of the Royal Society of London* 1952, B 237:37–72.

65   An embryo of a human . . . Gilbert, *Developmental Biology;* and N. H. Shubin and P. A. Alberch, "A morphogenetic approach to the origin and basic organization of the tetrapod limb," *Evolutionary Biology* 1986, 20:319–387.

66   Some researchers looked for morphogenetic molecules . . . S. A. Newman and H. L. Frisch, "Dynamics of skeletal pattern formation in developing chick limb," *Science* 1979, 205:662–668.

66   When a group of biologists led by George Oster . . . The mechanochemical model is explored in papers including G. F. Oster, J. D. Murray, and A. K. Harris, "Mechanical aspects of mesenchymal morphogenesis," *Journal of Embryology and Experimental Morphology* 1983, 78:83–125; and J. D. Murray and P. K. Maini, "Mechanochemical models for generating biological pattern and form in development," *Physics Reports* 1988, 171:59–84.

67   While working at Harvard in 1983 . . . P. Alberch and E. A. Gale, "Size dependence during the development of the amphibian foot. Colchicine-induced digital loss and reduction," *Journal of Embryology and Experimental Morphology* 1983, 76:177–197.

67   The genes of each breed . . . P. Alberch, "Developmental constraints: Why St. Bernards often have an extra digit and poodles never do," *American Naturalist* 1985, 126:430–433; and "Possible Dogs," *Natural History*, December 1986.

67   The harshest refutation came from experiments in the 1970s . . . J. R. Hinchliffe, "Developmental approaches to the problem of transformation of limb structure in evolution," in *Developmental Patterning of the Vertebrate Limb*, ed. J. R. Hinchliffe et al., pp. 313–323 (New York: Plenum Press, 1991).

69   They could sketch out the growth with a set of symbols . . . Shubin and Alberch, "A morphogenetic approach."

71   You could grow five fingers . . . M. I. Coates and J. A. Clack, "Polydactyly in the earliest known tetrapod limbs," *Nature* 1990, 347:66–69.

72   In 1990 English biologists . . . L. Wolpert and A. Hornbruch, "Double anterior chick limb buds and models for cartilage rudiment specification," *Development* 1990, 109:961–966.

72   One set of freaks . . . W. Bateson, *Materials for the Study of Variation Treated with Especial Regard to Discontinuity in the Origin of Species* (London: Macmillan, 1894).

72 It was only in the 1980s that geneticists . . . For recent reviews of homeotic gene studies, see E. M. De Robertis, "Homeotic genes and the evolution of body plans," in *Evolution and the Molecular Revolution,* ed. C. R. Marshall and J. W. Schopf, pp. 109–124 (Sudbury, Mass.: Jones and Bartlett, 1996); and R. A. Raff, *The Shape of Life: Genes, Development, and the Evolution of Form* (Chicago: University of Chicago Press, 1996).

74 A fruit fly has only one set, but jawless vertebrates like the lamprey have three, and most jawed fish and tetrapods have four or five . . . F. Ruddle, "Vertebrate genome evolution," *Genomics,* 1997 (in preparation).

74 The duplication of an entire set . . . S. Ohno, *Evolution by Gene Duplication* (Heidelberg: Springer-Verlag, 1970).

74 Immune cells in many animals . . . P. Jollés, ed., *Lysozymes: Model Enzymes in Biochemistry and Biology* (Basel, Switzerland: Birkhauser Verlag, 1996).

75 the duplications of homeotic genes . . . P. W. H. Holland et al., "Gene duplications and the origins of vertebrate development," *Development* 1994 Supplement, 125–133.

75 A region of the limb bud . . . R. D. Riddle et al., "Sonic Hedgehog mediates the polarizing activity of the ZPA," *Cell* 1993, 75:1401–1416; L. Niswander et al., "A positive feedback loop coordinates growth and patterning in the vertebrate limb," *Nature* 1994, 371:609–612; C. Tabin, "The initiation of the limb bud: Growth factors, *Hox* genes, and retinoids," *Cell* 1995, 80:671–674.

75 When scientists first detected Hox genes . . . J.-C. Izisua-Belmonte et al., "Expression of the homeobox *Hox*-4 genes and the specification of position in chick wing development," *Nature* 350:585–589; and C. Tabin, "Isolation of potential vertebrate limb-identity genes," *Development* 1989, 105:813–820.

76 it was the curve of Shubin's digital arch . . . M. I. Coates, "New palaeontological contributions to limb ontgeny and phylogeny," in *Developmental Patterning of the Vertebrate Limb,* ed. J. R. Hinchliffe et al., pp. 325–337 (New York: Plenum Press, 1991).

76 Inspired by Coates, in 1995 Swiss biologists . . . P. Sordino, F. van der Hoeven, and D. Duboule, "*Hox* gene expression in teleost fins and the origin of vertebrate digits," *Nature* 1995, 375:678–681.

77 The species they chose . . . N. Holder and A. McMahon, "Genes from zebrafish screens," *Nature* 1996, 384:515–516.

77   These results agreed with the predictions . . . M. I. Coates, "Fish fins or tetrapod limbs—a simple twist of fate?" *Current Biology* 1995, 5:844–848; and P. Thorogood, "The development of the teleost fin and implications for our understanding of tetrapod limb evolution," in *Developmental Patterning of the Vertebrate Limb,* ed. J. R. Hinchliffe et al., pp. 347–354 (New York: Plenum Press, 1991).

77   In 1996 an American group of geneticists . . . C. E. Nelson et al., "Analysis of *Hox* gene expression in the chick limb bud," *Development* 1996, 122:1449–1466; and N. Shubin, C. Tabin, and S. Carroll, "Fossils, genes, and the evolution of animal limbs," *Nature* 1997 388:639–648.

79   The fossil belonged . . . T. Daeschler, "Preliminary description of the pectoral girdle and fin from a new cf. *Sauripterus* specimen," *Journal of Vertebrate Paleontology* 1996, 16:29A; and Shubin et al., "Fossils, genes, and the evolution of animal limbs."

80   When William Bateson . . . E. Mayr, "The emergence of evolutionary novelties," in *Evolution after Darwin,* ed. Sol Tax (Chicago: University of Chicago Press, 1960).

80   "we read his scheme . . ." Quoted in Mayr, *Growth of Biological Thought,* p. 547.

80   A leading advocate of this shifting paradigm is Guenter Wagner . . . Wagner discusses evolutionary novelties in G. B. Müller and G. P. Wagner, "Novelty in evolution: Restructuring the concept," *Annual Review of Ecology and Systematics* 1991, 22:229–256; and G. P. Wagner, "The origin of morphological characters and the biological basis of homology," *Evolution* 1989, 43:1157–1171. Mechanisms for generating novelty are discussed in G. B. Müller, "Developmental mechanisms at the origin of morphological novelty: A side-effect hypothesis," in *Evolutionary Innovations,* ed. M. Nitecki, pp. 99–130 (Chicago: University of Chicago Press, 1990).

81   On birds, the tibia . . . G. B. Müller and J. Streicher, "Ontogeny of the syndesmosis tibiofibularis and the evolution of the bird hindlimb: A caenogenetic feature triggers phenotypic novelty," *Anatomy and Embryology* 1989, 179:327–339.

82   When many rodents forage . . . P. Brylski and B. K. Hall, "Ontogeny of macroevolutionary phenotype: The external cheek pouches of geomyoid rodents," *Evolution* 1988, 42:391–395.

83   As for the tetrapod limb . . . See Shubin et al., "Fossils, genes, and the evolution of animal and limbs"; and F. van der Hoeven, J. Zakany, and D. Duboule, "Gene

transposition in the *HoxD* complex reveals a hierarchy of regulatory controls," *Cell* 1996, 85:1025–1035.

84   In 1991 a freak freeze . . . N. H. Shubin and D. Wake, "Phylogeny, variation, and morphological integration," *American Zoologist* 1996, 36:51–60. Two other important papers on constraints are J. Maynard Smith et al., "Developmental constraints and evolution," *Quarterly Review of Biology* 1985, 60:265–287; and D. B. Wake, "Homoplasy: The result of natural selection or evidence of design limitations?" *American Naturalist* 1991, 138:543–567.

85   Some still argue that a Turing pattern . . . S. A. Newman, "Sticky fingers: *Hox* genes and cell adhesion in vertebrate limb development," *BioEssays* 1996, 18:171–174.

CHAPTER 4. DARWIN'S SAPLINGS
Interviews with Per Ahlberg, Elizabeth Brainerd, Jenny Clack, Michael Coates, Ted Daeschler, James Edwards, Colleen Farmer, Stephen Heard, George Lauder, Brian Moore, Kevin Padian, Joseph Skulan, Stuart Sumida, and Keith Thomson.

86   Sound is a gentle shift . . . Clack's papers on the evolution of the ear include "Discovery of the earliest known stapes," *Nature* 1989, 342:425–427; "Nos ancêtres respiraient-ils par les oreilles?" *La Recherche* 1990, 21:770–771; "The stapes of *Acanthostega gunnari* and the role of the stapes in early tetrapods," in *The Evolutionary Biology of Hearing*, ed. D. Webster, R. Fay, and A. Popper, pp. 405–420 (New York: Springer-Verlag, 1992); "Homologies in the fossil record: The middle ear as a test case," *Acta Biotheoretica* 1993, 41:391–409; and "Earliest known tetrapod braincase and the evolution of the stapes and fenestra ovalis," *Nature* 1994, 369:392–394. Also worth reading is R. I. Carroll, "The hyomandibular as a supporting element in the skull of primitive tetrapods," in *The Terrestrial Environment and the Origin of Land Vertebrates*, ed. A. L. Panchen, pp. 255–292 (London: Academic Press, 1980).

88   This choreography of changes caught the attention of Keith Thomson . . . Thomson initially proposed correlated progression in "The evolution of the tetrapod middle ear in the rhipidistian-amphibian transition," *American Zoologist* 1966, 6:379–397; and returned to it in "Fisher's microscope, or the gradualist's dilemma," *American Scientist* 1988, 76:500–502; "The origin of the tetrapods," *American Journal of Science* 1993, 293–A:33–62; and "Macroevolution: The morphological problem," *American Zoologist* 1996, 31:106–112.

90   Not far from the stapes, Clack and Coates discovered a strut of bone . . . M. I. Coates and J. A. Clack, "Fish-like gills and breathing in the earliest known tetrapods," *Nature* 1991, 352:234–236.

90   The longer Clack and Coates looked . . . See the two papers Clack and Coates
     contributed to the volume *Studies on Early Vertebrates (7th International Sympo-
     sium, Miguasha Parc, Quebec), Bulletin du Musée Nationale de L'Histoire Naturelle,*
     Paris 1995, 17(C): "*Acanthostega*—a primitive aquatic tetrapod?" pp. 359–72,
     and "Romer's Gap—tetrapod origins and terrestriality," pp. 373–388. Also see
     Coates, "The Devonian tetrapod *Acanthostega*."

91   An answer of sorts . . . T. W. Pietsch and D. B. Grobecker, *Frogfishes of the World:
     Systematics, Zoogeography, and Behavioral Ecology* (Stanford: Stanford University
     Press, 1987); and T. W. Pietsch, "Louis Renard's fanciful fishes," *Natural History*
     1984, 93:58–67.

92   "a sort of arms . . ." Pietsch and Grobecker, *Frogfishes,* p. 30.

92   "It followed me . . ." Pietsch, "Louis Renard's fanciful fishes," p. 64.

92   One species that lives in the vast forests of kelp . . . R. F. Sisson, "Adrift on a raft
     of sargassum," *National Geographic* 1976, 149:188–199.

92   From time to time scientists have pointed out . . . J. L. Edwards, "Two perspec-
     tives on the evolution of the tetrapod limb," *American Zoologist* 1989,
     29:235–254.

94   He has now tentatively reconstructed the animal . . . Ahlberg describes *Elginer-
     peton* in "Tetrapod or near-tetrapod fossils from the Upper Devonian of Scot-
     land," *Nature* 1991, 354:298–301; "*Elginerpeton pancheni* and the earliest
     tetrapod clade," *Nature* 1995, 373:420–425; and "Spare parts for *Elginerpeton?*
     The postcranial steam tetrapod remains from Scat Craig, Morayshire, Scotland,"
     Zoological Journal of the Linnean Society, in press.

95   a German entomologist named Willi Hennig . . . Among the many reviews of
     cladistics are D. L. Hull, *Science as a Process* (Chicago: University of Chicago
     Press, 1988); G. Nelson, "Homology and systematics," in *Homology: The Hierar-
     chical Basis of Comparative Biology,* ed. B. K. Hall, pp. 101–149 (New York: Aca-
     demic Press, 1994); and M. Ridley, *Evolution and Classification: The Reformation
     of Cladism* (London: Longman, 1986).

98   Some cladists even claimed (wrongly) . . . The danger of ignoring fossils is illus-
     trated in J. A. Gauthier, A. G. Kluge, and T. Rowe, "Amniote phylogeny and the
     importance of fossils," *Cladistics* 1988, 4:105–209.

98   And yet the information that we have in fossils and living animals . . . G. F. Engelmann and E. O. Wiley, "The place of ancestor-descendant relationship in phylogeny reconstruction," *Systematic Zoology* 1977, 26:1–11.

100   He keyed in information . . . The tree and much of its interpretation are drawn from Coates, "The Devonian tetrapod *Acanthostega*." For a cladogram that tells a somewhat different story of the origins of amphibians and amniotes, see M. Laurin and R. R. Reisz, "A new perspective on tetrapod phylogeny," in *Amniote Origins: Completing the Transition to Land*, ed. S. S. Sumida and K. L. M. Martin (New York: Academic Press, 1997).

100   He thought that fishes . . . S. J. Gould, *Eight Little Piggies* (New York: Norton, 1996). The revised cladogram is based on K. F. Liem, "Forms and function of lungs: The evolution of air-breathing mechanisms," *American Zoologist* 1988, 28:739–759.

102   On the other hand, as Colleen Farmer of Brown University has pointed out . . . C. Farmer, "Did lungs and the intracardiac shunt evolve to oxygenate the heart in vertebrates?" *Paleobiology* 1997, 23:358–372.

102   Cladistics allows researchers to reconstruct . . . E. L. Brainerd and J. S. Ditelberg, "Lung ventilation in salamanders and the evolution of vertebrate air-breathing mechanisms," *Biological Journal of the Linnean Society* 1993, 49:163–183.

104   One of these was an animal called *Panderichthys* . . . E. I. Vorobyeva and R. Hinchliffe, "From fin to limbs: Developmental perspectives on paleontological and morphological evidence," *Evolutionary Biology* 1996, 29:263–311.

106   the Russian animal known as *Tulerpeton* . . . O. A. Lebedev, "*Tulerpeton*, l'animal à six doigts," *La Recherche* 1990, 21:1274–1275; and O. A. Lebedev and M. I. Coates, "The postcranial skeleton of the Devonian tetrapod *Tulerpeton curtum* Lebedev," *Zoological Journal of the Linnean Society* 1995, 114:307–348.

108   The first amphibians surged into more than a dozen major lineages . . . P. E. Ahlberg and A. R. Milner, "The origin and early diversification of tetrapods," *Nature* 1994, 368:507–514.

108   Living amphibians . . . R. C. Stebbins and N. W. Cohen, *A Natural History of Amphibians* (Princeton: Princeton University Press, 1995); and W. E. Duellman and L. Trueb, *Biology of Amphibians* (New York: McGraw Hill, 1986).

109   The forerunners of amniotes . . . See Sumida and Martin, *Amniotes Origins*.

110   Alfred Romer was of this opinion . . . A. S. Romer, "Origin of the amniote egg," *Scientific Monthly* 1957, 82:57–63; and "Tetrapod limbs and early tetrapod life," *Evolution* 1958, 12:365–369.

110   In fact, some basic physics shows . . . J. Skulan, "Alternative scenarios for the origin of the amniote egg," *Journal of Vertebrate Paleontology* 1995, 15:54A.

111   Biologists often refer to these secrets to success as "key innovations" . . . S. B. Heard and D. L. Hauser, "Key evolutionary innovations and their ecological mechanisms," *Historical Biology* 1995, 10:151–173.

112   It may just happen to live . . . J. Cracraft, "The origin of evolutionary novelty: Pattern and process at different hierarchical levels," in *Evolutionary Innovations*, ed. M. Nitecki, pp. 21–43 (Chicago: University of Chicago Press, 1990).

113   In 1996 Robert Reisz and fellow scientists . . . See the following papers in *Sixth North American Paleontological Convention, Paleontological Society Special Publication* Number 8, 1996: B. R. Moore and D. R. Brooks, "A comparative analysis of herbivory and amniote diversification in modern terrestrial ecosystems," p. 80; and R. R. Reisz and B. R. Moore, "Exploring the 'egg/plant question': What best explains the patterns of amniote diversification?" p. 322.

114   As with the origin of tetrapods . . . M. S. Lee, "Correlated progression and the early evolution of turtles," *Nature* 1996, 379:812–815; and P. M. Barrett, "Correlated progression and the evolution of herbivory in non-avian dinosaurs," *Journal of Vertebrate Paleontology* 1997, 17:31A.

115   Among lizards, for example . . . F. H. Pough, "Lizard energetics and diet," *Ecology* 1973, 54:837–844.

CHAPTER 5. THE MIND AT SEA

Interviews with Ted Cranford, Jim Darling, Frank Fish, Louis Herman, John Heyning, Harry Jerison, Lori Marino, Bill McLellan, Ann Pabst, and Terrie Williams.

118   There are seventy-nine known living species . . . See R. Ellis, *The Book of Whales* (New York: Knopf, 1980), and *Dolphins and Porpoises* (New York: Knopf, 1982).

119   In order to be a graceful swimmer . . . For a review of dolphin hydrodynamics, see F. E. Fish, "Dolphin swimming–a review," *Mammal Review* 1991, 21:181–195.

120   In 1936 Sir James Gray . . . J. Gray, "Studies in animal locomotion. VI. The propulsive powers of the dolphin," *Journal of Experimental Biology* 1936, 13:192–199.

121 Physiologists have found that a dolphin has the cheapest cost of transport . . . T. M. Williams et al., "The physiology of bottlenose dolphins *(Tursiops truncatus):* Heart rate, metabolic rate, and plasma lactate concentration during exercise," *Journal of Experimental Biology* 1993, 179:31–46; and T. M. Williams, "The evolution of cost-efficient swimming in marine mammals: optimizing energetics," *Philosophical Transactions of the Royal Society of London* 1998, B (in press).

122 It may look like they are doing it for pleasure . . . T. M. Williams et al., "Travel at low energetic cost by swimming and wave-riding bottlenose dolphins," *Nature* 1992, 355:821–823.

123 A male dolphin preserves his fertility . . . S. A. Rommel et al., "Anatomical evidence for a countercurrent heat exchanger associated with dolphin testes," *Anatomical Record* 1992, 232:150–156; S. A. Rommel et al., "Temperature regulation of the testes of the bottlenose dolphin *(Tursiops truncatus):* Evidence from colonic temperatures," *Journal of Comparative Physiology* 1994, B 164:130–134; and D. A. Pabst, "Thermoregulation of the intra-abdominal testes of the bottlenose dolphin *(Tursiops truncatus)* during exercise," *Journal of Experimental Biology* 1995, 198:221–226.

123 Female dolphins need just as badly . . . S. A. Rommel, D. A. Pabst, and W. A. McLellan, "Functional morphology of the vascular plexuses associated with the cetacean uterus," *Anatomical Record* 1992, 237:538–546.

124 Ann Pabst, a biologist at the University of North Carolina . . . D. A. Pabst, "Intramuscular morphology and tendon geometry of the epaxial swimming muscles of dolphins," *Journal of Zoology, London* 1993, 230:159–176; and "Morphology of the subdermal connective tissue sheath of dolphins: A new fibre-wound, thin-walled, pressurized cylinder model for swimming vertebrates," *Journal of Zoology, London* 1996, 238:35–52.

126 The best guide through the cogs and gears . . . T. W. Cranford, M. Amundin, and K. S. Norris, "Functional morphology and homology in the odontocete nasal complex: Implications for sound generation," *Journal of Morphology* 1996, 228:223–285.

130 Researchers have been able to gauge the perceptions of dolphins . . . W. Au, *The Sonar of Dolphins* (New York: Springer-Verlag, 1993).

130 Researchers have had dolphins look at intricate constructions of plastic pipes . . . A. A. Pack and L. M. Herman, "Sensory integration in the bottlenose dolphin in immediate recognition of complex shapes across the senses of echolocation and vision," *Journal of the Acoustic Society of America* 1995, Part 1, 98:722–732.

130   At the University of Hawaii . . . Louis Herman and his colleagues have produced
a long string of papers on their dolphin work. See, for example, L. M. Herman,
D. G. Richards, and J. P. Wolz, "Comprehension of sentences by bottlenosed dolphins," *Cognition* 1984, 16:129–219; and "Representational and conceptual
skills of dolphins," in *Language and Communication: Comparative Perspectives,* ed.
H. L. Roitblat, L. M. Herman, and P. E. Nachtigall, pp. 403–422 (Hillsdale,
N.J.: Lawrence Erlbaum, 1993).

131   they know when it was being violated . . . L. M. Herman, S. A. Kuczaj, and
M. D. Holder, "Responses to anomalous gestural sequences by a language-trained
dolphin: Evidence of processing of semantic relations and syntactic information,"
Journal of Experimental Psychology, General 1993, 122:184–194.

132   A researcher at Emory University in Georgia named Lori Marino . . . L. A.
Marino, D. Reiss, and G. G. Gallup, "Mirror self-recognition in bottlenose dol-
phins: Implications for comparative investigations of highly dissimilar species,"
in *Self-awareness in Animals and Humans: Development Perspectives,* ed. S. Parker,
R. Mitchell, and M. Boccia, pp. 380–391 (New York: Cambridge University
Press, 1994).

133   Harry Jerison . . . H. J. Jerison, "The perceptual worlds of dolphins," in *Dolphin
Cognition and Behavior: A Comparative Approach,* ed. R. J. Schusterman, J. A.
Thomas, and F. G. Wood, pp. 141–166 (Hillsdale, N. J.: Lawrence Erlbaum,
1986).

CHAPTER 6. THE EQUATION OF A WHALE
Interviews with Phillip Gingerich, Maureen O'Leary, Martin Rudwick, Phillip
Sloan, Hans Thewissen.

135   he "rebuilt, like Cadmus, cities from a tooth." Quoted from Appel, *The Cuvier-
Geoffroy Debate,* p. 190.

135   The Baron of Fossils earned his reputation . . . This episode is drawn from G.
Cuvier, *Recherches sur les ossements fossiles* (Paris: d'Ocagne, 1836); with additional
information from W. Coleman, *Georges Cuvier, Zoologist* (Cambridge: Harvard
University Press, 1964); and M. J. S. Rudwick, *Georges Cuvier, Fossil Bones, and
Geological Catastrophes* (Chicago: University of Chicago Press, 1997).

136   "I stopped my work on the teeth . . ." Cuvier, *Recherches,* p. 8.

136   "Every organized creature forms a whole . . ." quoted in Coleman, *Georges Cuvier,*
p. 119.

136   "This operation was done . . ." Cuvier, *Recherches,* p. 10.

137   "There is no science that cannot become almost geometrical . . ." Cuvier, *Recherches,* p. 16.

138   The letter was from a Louisiana man . . . R. D. Harlan, "Notices of fossil bones found in the Tertiary Formation of the State of Louisiana," *Transactions of the American Philosophical Society* 1834, 4:397–403; *Medical and Physical Researches: Or Original Memoirs in Medicine, Surgery, Physiology, Geology, Zoology, and Comparative Anatomy* (Philadelphia: Lydia R. Bailey, 1835); and "A Letter from Dr. Harlan, addressed to the president, on the discovery of the remains of the *Basilosaurus* or *Zeuglodon,*" *Transactions of the Geological Society of London,* series 2, 1841, 6:67–68.

138   "A scientific memoir . . ." Harlan, "Notice of fossil bones," p. 397.

139   "If future discoveries . . ." Harlan, "Notice of fossil bones," p. 403.

139   When word got to Europe . . . A. Desmond, *The Politics of Evolution; Morphology, Medicine, and Reform in Radical London* (Chicago: University of Chicago Press, 1989); and P. A. Gerstner, "Vertebrate paleontology: An early nineteenth-century transatlantic science," *Journal of the History of Biology* 1970, 3:137–148.

140   A few weeks later at the meeting . . . R. A. Owen, "Observations on the teeth of the *Zeuglodon* (*Basilosaurus*) of Doctor Harlan," *Proceedings of the Geological Society of London* 1839, 3:24–28; and "Observations on the *Basilosaurus* of Dr. Harlan (*Zeuglodon cetoides,* Owen)," *Transactions of the Geological Society of London,* series 2, 1841, 6:69–80.

140   "one of the most extraordinary . . ." R. A. Owen, "Observations on the *Basilosaurus,*" p. 79.

141   The owner of the museum was a German immigrant named Albert Koch . . . R. Ellis, *Monsters of the Sea* (New York: Doubleday, 1995); A. C. Koch, *Journey through a part of the United States of North America in the Years 1844 to 1846,* trans. and ed. Ernst A. Stadler (Carbondale: Southern Illinois University Press, 1972); and K. V. W. Palmer, "Tales of ancient whales," *Nature Magazine* 1942, 35:213.

141   "the sovereign masterpiece . . ." 1842 poster, courtesy of Michael Coates.

143   In later years paleontologists . . . M. Uhen, "What is *Pontogeneus brachyspondylus?*" *Journal of Vertebrate Paleontology* 1997, 17:82A.

143   A gentleman from Old Washington Courthouse . . . J. Wynan, "A communication from Professor Jeffries Wynan, on the subject of the fossil recently exhibited in New York as that of a sea-serpent under the name of *Hydrarchos sillimani,*" *Proceedings of the Boston Society of Natural History* 1845, 2:65–68.

144   The new skeleton prompted people to ask Richard Owen . . . Rupke, *Richard Owen,* and R. A. Owen, "The great sea-serpent," *Annals and Magazine of Natural History,* series 2, 1848, 2:317–322.

144   After Koch's discoveries . . . A. R. Kellogg, "The history of whales—their adaptation to life in the water," *Quarterly Review of Biology* 1928, 3:29–76 and 174–208; and "A review of the Archaeoceti," *Carnegie Institute of Washington Publication* 1936, 482:1–366.

144   "In North America . . ." Charles Darwin, *On the Origin of Species.* (London: Penguin Books, 1985), p. 215.

145   "Mr. Darwin has . . ." Alvar Ellegård, *Darwin and the General Reader* (Chicago: University of Chicago Press, 1990), p. 240.

145   Flower was twenty-eight . . . C. J. Cornish, *Sir William Henry Flower: A Personal Memoir* (London: Macmillan, 1904); and R. Lydekker, *Sir William Flower* (London: J. M. Dent, 1906).

146   "There is no epoch-making discovery . . ." Lydekker, *Sir William Flower,* p. 175.

146   And toward the close of his career . . . W. H. Flower, "On whales, past and present, and their probable origin," *Notices of the Proceedings of the Royal Institution of Great Britain* 1883, 10:360–376.

146   "Scarcely anywhere in the animal kingdom . . ." W. H. Flower, "On whales," p. 1.

148   the report of whalers from Vancouver Island who killed a humpback female . . . R. A. Andrews, "A remarkable case of external hind limbs in a humpback whale," *American Museum Novitates* 1921, 9:1–6.

148   No, the real trouble . . . Lydekker, *Sir William Flower.*

149   The earliest synapsids . . . Much of the latest thinking on the evolution of mammals is summarized in R. I. Carroll, *Vertebrate Paleontology and Evolution* (San Francisco: Freeman, 1988); and T. S. Kemp, *Mammal-Like Reptiles and the Origin of Mammals* (London: Academic Press, 1982).

150   Now the muscles they used for breathing no longer had to help during walking . . . D. R. Carrier, "The evolution of locomotor stamina in tetrapods: Circumventing a mechanical constraint," *Paleobiology* 1987, 13:326–341; and "Conflict in the hypaxial musculo-skeletal system: Documenting an evolutionary conflict," *American Zoologist* 1991, 31:644–654. See also D. M. Bramble and F. A. Jenkins, "Structural and functional integration across the reptile-mammal boundary: The locomotory system," in *Complex Function: Integration and Evolution in Vertebrates*, ed. D. B. Wake and G. Roth, pp. 133–146 (New York: John Wiley & Sons, 1989).

150   They could pump more molecules . . . J. Ruben, "The evolution of endothermy in mammals and birds: From physiology to fossils," *Annual Review of Physiology* 1995, 57:69–95.

150   The earliest synapsids probably heard the world . . . See the work of J. A. Hopson, including "The origin of the mammalian middle ear," *American Zoologist* 1966, 6:437–450; and "Synapsid evolution and the radiation of non-eutherian mammals," in *Major Features of Vertebrate Evolution: Short Courses in Paleontology No. 7*, ed. D. R. Prothero and R. M. Schoch, pp. 238–270 (Knoxville, Tenn.: Paleontological Society, 1994).

151   When they grew as embryos . . . T. Rowe, "Coevolution of the mammalian middle ear and neocortex," *Science* 1996, 273:651–654.

151   These new signals entered a brain that was now awash with signals . . . H. J. Jerison, *Evolution of the Brain and Intelligence* (New York: Academic Press, 1973); and *Brain Size and the Evolution of Mind*, 59th James Arthur Lecture on the Evolution of the Human Brain (New York: American Museum of Natural History, 1991).

151   Paleontologists still debate . . . J. D. Archibald, *Dinosaur Extinction and the End of an Era: What the Fossils Say* (New York: Columbia University Press, 1996).

152   In 1950 biologists at Rutgers University . . . A. Boyden and D. Gemeroy, "The relative position of the Cetacea among the orders of Mammalia as indicated by precipitin tests," *Zoologica* 1950, 35:145–151.

153   In 1966 Leigh Van Valen . . . L. Van Valen, "Deltatheridia, a new order of mammals," *Bulletin of the American Museum of Natural History* 1966, 132:1–126.

155   One of the first descriptions of a mesonychid . . . E. D. Cope, "The Creodonta," *American Naturalist* 1884, 18:255–267.

156 O'Leary made a special study . . . M. A. O'Leary and K. D. Rose, "Postcranial skeleton of the early Eocene mesonychid *Pachyaena* (Mammalia: Mesonychia)," *Journal of Vertebrate Paleontology* 1995, 15:401–430. Other mesonychid papers include F. S. Szalay, "Origin and evolution of function of the mesonychid condylarth feeding mechanism," *Evolution* 1969, 23:703–720; and X. Zhou, W. J. Sanders, and P. D. Gingerich, "Functional and behavioral implications of vertebral structure in *Pachyaena ossifraga* (Mammalia, Mesonychia)," *Contributions from the Museum of Paleontology, the University of Michigan* 1992, 28:289–319.

CHAPTER 7. ALONG THE TETHYAN SHORES
Interviews with Phillip Gingerich, Jean-Louis Hartenberger, Charles Marshall, James Reilly, Melanie Stiassny, and Mark Uhen.

162 The rocks were not spectacular . . . P. D. Gingerich, "A small collection of fossil vertebrates from the Middle Eocene Kuldana and Kohat formations of Punjab (Pakistan)," *Contributions from the Museum of Paleontology* 1977, 24:190–203.

163 But two months after his visit . . . R. M. West, "Middle Eocene large mammal assemblage with Tethyan affinities, Ganda Kas region, Pakistan," *Journal of Paleontology* 1980, 54:508–533.

164 Meanwhile Neil Wells . . . N. A. Wells, "Transient streams in sand-poor redbeds: Early-Middle Eocene Kuldana Formation of northern Pakistan," *Special Publications of the International Association of Sedimentologists* 1983, 6:393–403.

164 One of Gingerich's fellow rock smashers . . . P. D. Gingerich and D. E. Russell, "*Pakicetus inachus*, a new archaeocete (Mammalia, Cetacea) from the Early-Middle Eocene Kuldana Formation of Kohat (Pakistan)," *Contributions from the Museum of Paleontology* 1981, 25:235–246; P. D. Gingerich et al., "Origin of whales in epicontinental remnant seas: New evidence from the early Eocene of Pakistan," *Science* 1983, 220:403–406; and P. D. Gingerich, "The whales of Tethys," *Natural History* April 1994, 86–88.

165 They were certainly important . . . P. D. Gingerich and D. E. Russell, "Dentition of Early Eocene *Pakicetus* (Mammalia, Cetacea)," *Contributions from the Museum of Paleontology* 1990, 28:1–20.

166 A century earlier to the year . . . P. D. Gingerich, "Marine mammals (Cetacea and Sirenia) from the Eocene of Gebel Mokattam and Fayum, Egypt: Stratigraphy, age, and paleoenvironments," *University of Michigan Papers on Paleontology* 1992, 30:1–84.

166 "Very often a fragment . . ." P. E. P. Deraniyagala, "Some scientific results of two visits to Africa," *Spolia Zeylanica* 1948, 25:1–42; see p. 2.

168 His search for hips . . . P. D. Gingerich, B. H. Smith, and E. L. Simons, "Hind limbs of Eocene *Basilosaurus:* Evidence of feet in whales," *Science* 1990, 249:154–157.

170 To Darwin, such shrinking vestiges . . . Mayr, *The Growth of Biological Thought.*

170 We now have a good grip on how evolution's eraser works . . . D. W. Fong, R. C. Kane, and D. C. Culver, "Vestigialization and loss of nonfunctional characters," *Annual Review of Ecology and Systematics* 1995, 26:249–268; and C. R. Marshall, E. C. Raff, and R. A. Raff, "Dollo's law and the death and resurrection of genes," *Proceedings of the National Academy of Sciences* 1994, 91:12283–12287.

171 Take the gristly leg . . . Atavisms are surveyed in B. K. Hall, "Developmental mechanisms underlying the formation of atavisms," *Biological Reviews of the Cambridge Philosophical Society* 1984, 59:89–124.

171 thick mats of hair . . . L. E. Figuero et al., "Mapping of the congenital generalized hypertichosis locus to chromosome Xq24–q271," *Nature Genetics* 1995, 10:202–207.

171 In 1980 scientists took some of this gum tissue . . . E. J. Kollar and C. Fisher, "Tooth induction in chick epithelium: Expression of quiescent genes for enamel synthesis," *Science* 1980, 207:993–995.

172 In a recent experiment, geneticists plucked a gene . . . J. R. Brown et al., "A defect in nurturing in mice lacking the immediate early gene *fosB*," *Cell* 1996, 86:297–309.

172 Some researchers have reported little luck . . . Marshall et al., "Dollo's law."

172 In the early 1990s . . . K. K. Smith and R. A. Schneider, "Gene knockouts and the mammalian first arch: Evolutionary reversals or developmental disruptions?" *American Zoologist* 1996, 36:56A.

172 Consider a gene *A* . . . Marshall et al., "Dollo's law."

173 As animals move into darkness . . . J. N. Lythgoe, *The Ecology of Vision* (Oxford: Clarendon Press, 1979).

174   One particularly desperate example . . . C. Zimmer, "The Light at the Bottom of the Sea," *Discover*, November 1996.

174   While studying Australian skinks . . . C. Gans, "Tetrapod limblessness: Evolution and functional corollaries," *American Zoologist* 1975, 15:455–467; and "Motor coordination factors in the transition from tetrapod to limblessness in lower vertebrates," in *Coordination of Motor Behavior*, Seminar Series/Society for Experimental Biology: 24, ed. B. M. H. Bush and F. Clarac, pp. 184–201 (New York: Cambridge University Press, 1985). See also R. Lande, "Evolutionary mechanisms of limb loss in tetrapods," *Evolution* 1978, 32:73–92.

175   Caecilians are amphibians . . . J. C. O'Reilly, D. A. Ritter, and D. R. Carrier, "Hydrosatic locomation in a limbless tetrapod," *Nature* 1997, 386:269–272.

175   Blind mole rats . . . J. M. Diamond, "Competition for brain space," *Nature* 1996, 382:756–766.

CHAPTER 8. WALKING TO SWIMMING
Interviews with Frank Fish, Phillip Gingerich, and Hans Thewissen.

180   Fish measured their body temperature . . . F. E. Fish, "Thermoregulation in the muskrat *(Ondatra zibethicus)*: The use of regional heterothermia," *Comparative Biochemistry and Physiology* 1979, 64A:391–397; and "Aerobic energetics of surface swimming in the muskrat *(Ondatra zibethicus)*," *Physiological Zoology* 1982, 55:180–189.

181   To make a careful study of their swimming . . . F. E. Fish, "Mechanics, power output, and efficiency of the swimming muskrat *(Ondatra zibethicus)*," *Journal of Experimental Biology* 1984, 110:183–201.

183   Step one would be a dog paddle . . . F. E. Fish, "Transitions from drag-based to lift-based propulsion in mammalian swimming," *American Zoologist* 1996, 36:628–641.

183   The North American opossum . . . F. E. Fish, "Comparison of swimming kinematics between terrestrial and semiaquatic opossums," *Journal of Mammalogy* 1993, 74:275–284.

184   They had to become, for a time, like otters . . . F. E. Fish, "Association of propulsive mode with behavior in river otters *(Lutra canadensis)*," *Journal of Mammalogy* 1994, 75:989–997.

184   a sea otter can swim 75 percent faster underwater . . . T. M. Williams, "Swimming by sea otters: Adaptations for low energetic cost locomotion," *Journal of Comparative Physiology* 1989, A 164:815–824.

188   David Krause . . . D. W. Krause and M. C. Maas, "The biogeographic origins of late Paleocene–early Eocene mammalian immigrants to the western interior of North America," in *Dawn of the Age of Mammals in the Northern Part of the Rocky Mountain Interior, North America*, Special Paper 243, ed. T. M. Bown and K. D. Rose, pp. 71–105 (Boulder, Colo.: Geological Society of America, 1990).

189   Arif, Aslan, and Thewissen . . . J. G. M. Thewissen, "Eocene marine mammals from the Himalayan foothills," *National Geographic Research & Exploration* 1993, 9:125–127.

190   The ears kept him . . . J. G. M. Thewissen and S. T. Hussein, "Origin of underwater hearing in whales," *Nature* 1993, 361:444–445.

194   Thewissen's creature, which he called *Ambulocetus* . . . J. G. M. Thewissen, S. T. Hussein, and M. Arif, "Fossil evidence for the origin of aquatic locomotion in archaeocete whales," *Science* 1994, 263:210–212.

196   Gingerich's whale, which he named *Rodhocetus* . . . P. D. Gingerich et al., "New whale from Pakistan and the origin of cetacean swimming," *Nature* 1994, 368:844–847.

CHAPTER 9. A VOYAGE OUT
Interviews with Larry Barnes, Ewan Fordyce, John Gatesy, Phillip Gingerich, John Heyning, David Irwin, Harry Jerison, Patrick Luckett, Lori Marino, Sharon Messenger, Michel Milinkovitch, Maureen O'Leary, Hans Thewissen, and Mark Uhen.

197   Like most stories about macroevolution . . . D. R. Prothero, "Mammalian evolution," in *Major Features of Vertebrate Evolution: Short Courses in Paleontology No. 7*, ed. D. R. Prothero and R. M. Schoch, pp. 238–270 (Knoxville, Tenn.: Paleontological Society, 1994).

198   Around the Tethys, however, one stock of mesonychids . . . This transformation is based in part on J. G. M. Thewissen, "Phylogenetic aspects of cetacean origins: A morphological perspective," *Journal of Mammalian Evolution* 1994, 2:157–184.

199   Gingerich thinks that only 2 million years . . . P. D. Gingerich and M. Uhen, "Likelihood estimation of the time of the origin of whales (Cetacea)," *Sixth North American Paleontological Convention, Paleontological Society Special Publication Number 8*, 1996:146.

199 In his mind, they were furry crocodiles . . . J. G. M. Thewissen, S. I. Madar, and S. T. Hussain, "*Ambulocetus natans,* an Eocene cetacean (Mammalia) from Pakistan," *Courier Forschungsinstitut Senkenberg* 1996, 191:1–86.

200 Gingerich has found at least three contemporaries . . . P. D. Gingerich, M. Arif, and W. C. Clyde, "New archaeocetes (Mammalia, Cetacea) from the Middle Eocene Domanda Formation of the Sulaiman Range, Punjab (Pakistan)," *Contributions from the Museum of Paleontology* 1995, 28:291–330.

201 They first measured the isotopes of oxygen . . . For living cetacean osmoregulation, see C. A. Hui, "Seawater consumption and water flux in the common dolphin *Delphinus delphis,*" *Physiological Zoology* 1981, 54:430–440. For archaeocete osmoregulation, see J. G. M. Thewissen et al., "Evolution of cetacean osmoregulation," *Nature* 1996, 381:379–380.

204 almost every bone in *Dorudon's* body is accounted for . . . M. D. Uhen, "*Dorudon atrox* (Mammalia, Cetacea): Form, function, and phylogenetic relationships of an archaeocete from the late Middle Eocene of Egypt," Ph.D. dissertation, University of Michigan, 1996.

205 A hint of what *Dorudon* ate . . . C. C. Swift and L. G. Barnes, "Stomach contents of *Basilosaurus cetoides:* Implications for the evolution of cetacean feeding behavior, and evidence for vertebrate fauna of epicontinental Eocene seas," *Sixth North American Paleontological Convention, Paleontological Society Special Publication Number 8,* 1996:380.

206 A possible solution appeared in 1966 . . . D. Emlong, "A new archaic cetacean from the Oligocene of northwest Oregon," *Bulletin of the Natural History Museum of the University of Oregon* 1966, 3:1–51.

206 But soon afterward Leigh Van Valen argued . . . L. Van Valen, "Monophyly or diphyly in the origin of whales," *Evolution* 1968, 22:37–41.

207 *Archeomysticetus* was small by the standards of today's baleens . . . L. G. Barnes and A. E. Sanders, "The transition from archaeocetes to mysticetes: Late Oligocene toothed mysticetes from near Charleston, South Carolina," *Sixth North American Paleontological Convention, Paleontological Society Special Publication Number 8,* 1996:24.

208 Other fossils suggest further evolution . . . R. E. Fordyce and L. G. Barnes, "The evolutionary history of whales and dolphins," *Annual Review of Earth and Planetary Sciences* 1994, 22:419–455.

208   The first true baleen whales may have been the product of a rapidly changing world . . . See papers by R. E. Fordyce, including "Whale evolution and Oligocene southern ocean environments," *Palaeogeography, Palaeoclimatology, Palaeoecology* 1980, 31:319–336; and "Cetacean evolution and Eocene/Oligocene environments," in *Eocene-Oligocene Climatic and Biotic Evolution*, ed. D. R. Prothero and W. A. Berggren, pp. 368–381 (Princeton: Princeton University Press, 1992). See also the earlier paper by J. H. Lipps and E. Mitchell, "Trophic model for the adaptive radiations and extinctions of pelagic marine mammals," *Paleobiology* 1976, 2:147–155.

209   The oldest toothed whale fossil . . . J. L. Goedert and L. G. Barnes, "The earliest known odontocete: A cetacean wtih agorophiid affinities from latest Eocene to earliest Oligocene rocks in Washington State." *Sixth North American Paleontological Convention, Paleontological Society Special Publication Number 8*, 1996:148.

209   Their common heritage with artiodactyls . . . F. R. Walther, *Communication and Expression in Hoofed Mammals* (Bloomington: Indiana University Press, 1984).

209   Their melon may have already existed as a nose plug . . . J. E. Heyning and J. G. Mead, "Evolution of the nasal anatomy of cetaceans," in *Sensory Abilities of Cetaceans*, ed. J. Thomas and R. Kastelein, pp. 67–79 (New York: Plenum Press, 1990).

210   Meanwhile, the nose was moving up toward the top of the head . . . H. A. Oelschläger, "Evolutionary morphology and acoustics in the dolphin skull," in *Sensory Abilities of Cetaceans*, ed. J. Thomas and R. Kastelein, pp. 67–79 (New York: Plenum Press, 1990).

210   The ears nudged their way up . . . D. R. Ketten, "The marine mammal ear: Specializations for aquatic audition and echolocation," in *The Evolutionary Biology of Hearing*, ed. D. Webster, R. Fay, and A. Popper, pp. 717–750 (New York: Springer-Verlag, 1992).

212   Several teams of geneticists have been comparing the genes of whales . . . The first hints of a hippo/cetacean clade appeared in D. Irwin et al., "Evolution of the cytochrome b gene of mammals," *Journal of Molecular Evolution* 1991, 32:128–144; and D. M. Irwin and U. Arnason, "Cytochrome b gene of marine mammals: Phylogeny and evolution," *Journal of Mammalian Evolution* 1994, 2:37–55. Other papers include D. Graur and D. Higgins, "Molecular evidence for the inclusion of cetaceans within the order Artiodactyla," *Molecular Biology and Evolution* 1994, 11:357–364; and M. Shimamura et al., "Molecular evidences from retroposons that whales form a clade within even-toed ungulates," *Nature* 1997, 388:666–670. Gatesy's tree is adapted from J. Gatesy, "More DNA

support for a Cetacea/Hippopotamidae clade: The blood-clotting protein gene gamma-fibrinogen," *Molecular Biology and Evolution* 1997, 14:537–543.

213   In the early 1990s Michel Milinkovitch . . . Milinkovitch first published his phylogeny in M. C. Milinkovitch, G. Orti, and A. Meyer, "Revised phylogeny of whales suggested by mitochondrial ribosomal DNA sequences," *Nature* 1993, 361:346–348. The latest extensions of his work can be found in M. C. Milinkovitch, "Molecular phylogeny of cetaceans prompts revision of morphological transformations," *Trends in Ecology and Evolution* 1995, 10(8):328–334; and M. C. Milinkovitch et al., "Effects of character weighting and species sampling on phylogeny reconstruction: A case study based on DNA sequence data in cetaceans," *Genetics* 1996, 144:1817–1833. (For a cladistic analysis of morphology that supports the traditional phylogeny of modern whales, see J. E. Heyning, "Sperm whale phylogeny revisited: Analysis of the morphological evidence," *Marine Mammal Science* 1997, in press.)

217   The work of Gatesy and Milinkovitch recently inspired Luckett . . . W. P. Luckett, "Molecular and morphological analysis of conflicting relationships among Cetacea and Artiodactyla (Mammalia)," in *Molecules and Morphology in Systematics* (Paris: Muséum National d'Histoire Naturelle, in press). Luckett previously used the same method in "Suprafamilial relationships within Marsupiala: Resolution and discordance from multidisciplinary data," *Journal of Mammalian Evolution* 1994, 2:255–283.

219   Jerison first had to redeem . . . See H. J. Jerison, *Evolution of the Brain and Intelligence* (New York: Academic Press, 1974); and "Animal intelligence as encephalization," *Philosophical Transactions of the Royal Society of London* 1985, B 308:21–35.

220   As Harvard anthropologist Terrence Deacon has noted . . . T. Deacon, *The Symbolic Species: The Coevolution of Language and the Brain* (New York: Norton, 1997).

221   Consider the selection . . . L. A. Marino, "A comparison of encephalization betwen odontocete cetaceans and anthropoid primates," *Brain, Behaviour, and Evolution* 1997 (in press).

222   Their skeletons show that they were lemurlike animals . . . R. F. Kay, C. Ross, and B. A. Williams, "Anthropoid origins," *Science* 1997, 275:797–804.

222   Evidence exists for several different evolutionary pressures . . . L. A. Marino, "Brain-behavior relationships in cetaceans and primates: Implications for the

evolution of complex intelligence," Ph.D. dissertation, State University of New York, Albany, 1995.

222   Another possibility arises . . . R. M. Dunbar, *Grooming, Gossip, and the Evolution of Language* (Cambridge: Harvard University Press, 1996).

223   Hominids may have been forming a new kind of society . . . See Deacon, *The Symbolic Species.*

224   odontocetes climbed up to their high EQs . . . H. J. Jerison, "Brain and intelligence in whales," in *Whales and Whaling,* Vol. 2, pp. 159–197 (Canberra: C. J. Thompson, Australian Commonwealth Government Printer, 1978).

224   Family units of sperm whales . . . L. Weilgart, H. Whitehead, and K. Payne, "A colossal convergence," *American Scientist* 1996, 84:278–287.

225   higher EQs correspond to bigger pod sizes . . . L. A. Marino, "What can dolphins tell us about primate evolution?" *Evolutionary Anthropolgy* 1996, 5:81–85.

CHAPTER 10. ON THE TRAILS OF MACROEVOLUTION
Interviews with Phillip Gingerich, Stephen Jay Gould, Jeffrey Levinton, and Kevin Padian.

227   "The period when the class of reptiles flourished . . . " R. A. Owen, "Report on British fossil reptiles, part 2," *Report of the British Association for the Advancement of Science* 1841:60–204. Quote from p. 94.

227   Sixteen years earlier, Geoffroy had studied the skull . . . Appel, *The Cuvier-Geoffroy Debate.*

230   In 1972 Stephen Jay Gould attacked the puzzle of varying evolutionary rates . . . S. J. Gould and N. Eldredge, "Punctuated equilibrium comes of age," *Nature* 1993, 366:223–227.

231   Meanwhile, some researchers who have been trying to measure evolution's natural pace in living animals . . . See, for example, D. N. Reznick et al., "Evaluation of the rate of evolution in natural populations of guppies *(Poecilia reticulata),*" *Science* 1997, 276:1934–1937.

231   To others stasis means . . . See, for example, Dawkins, *The Blind Watchmaker* (New York: W. W. Norton, 1986).

232   Alan Cheetham at the Smithsonian Institution . . . A. H. Cheetham, "Tempo of evolution in a Neogene bryozona: Rates of morphologic change within and across species boundaries," *Paleobiology* 1986, 12:190–202.

232   The many changes necessary to build a tetrapod . . . R. L. Carroll, "Revealing the patterns of macroevolution," *Nature* 1996, 381:19–20.

232   Mosasaurs descend from the same group . . . M. DeBraga and R. L. Carroll, "The origin of mosasaurs as a model of macroevolutionary patterns and process," *Evolutionary Biology* 1993, 27:245–323.

234   But as Rachel Collins of the University of Washington and Christine Janis of Brown University have pointed out . . . R. Collins and C. M. Janis, "Morphological constraints on tetrapod feeding mechanisms: Why were there no suspension-feeding marine reptiles?" in *Ancient Marine Reptiles,* ed. J. M. Callaway and E. L. Nicholls (New York: Academic Press, 1996).

235   The tree we now see . . . The tree of life is being redrawn at a furious pace. One of its newest incarnations can be found in N. R. Pace, "A molecular view of microbial diversity and the biosphere," *Science* 1997, 276:734–740. For more detail on the metazoan branches, inspect the excellent, ever-growing "Tree of Life" on the Internet (http://phylogeny.arizona.edu/tree/phylogeny.html).

SELECTED BIBLIOGRAPHY

At the Water's Edge is based on scientific papers and books, as well as visits, telephone interviews, and electronic mail exchanges with scientists. At the head of the notes to each chapter I've listed the researchers with whom I communicated during the course of my work. Because this is a book for a popular audience and not a scientific review, I have cited only those written sources most relevant to each part of the book. To try to offer a complete bibliography—even of limb development alone—would require a few too many acres of trees to be cut. Some general works of interest are listed below; consider them as a good starting place for curious souls.

Carroll, R. J. *Vertebrate Paleontology and Evolution.* San Francisco: Freeman, 1988.

———. *Patterns and Processes of Vertebrate Evolution.* New York: Cambridge University Press, 1997.

Cowen, R. *History of Life.* 2nd ed. Boston: Blackwell, 1995.

Ellis, R. *The Book of Whales.* New York: Knopf, 1980.

———. *Dolphins and Porpoises.* New York: Knopf, 1982.

Gilbert, W. *Developmental Biology.* 4th ed. Sunderland, Mass.: Sinauer, 1994.

Hall, B.K. *Evolutionary Developmental Biology.* New York: Chapman & Hall, 1992.

Little, Colin. *The Terrestrial Invasion: An Ecophysiological Approach to the Origins of Land Animals.* Cambridge: Cambridge University Press, 1990.

Long, J.A. *The Rise of Fishes: 500 Million Years of Evolution.* Baltimore: Johns Hopkins University Press, 1995.

Mayr, E. *The Growth of Biological Thought.* Cambridge: Harvard University Press, 1982.

Raff, R. A. *The Shape of Life: Genes, Development, and the Evolution of Form.* Chicago: University of Chicago Press, 1996.

Ridley, M. *Evolution.* Boston: Blackwell, 1993.

Slijper, E. J. *Whales.* New York: Basic Books, 1962.

Thomson, K. S. *Morphogenesis and Evolution.* Oxford: Oxford University Press, 1988.

ACKNOWLEDGMENTS

Thanks must go first to all the scientists who let me tag along on their field work, lodge myself in their offices for days on end, pester them on the phone, and overload their modems with electronic mail. Particular gratitude goes to Robert Carroll, Michael Coates, Kevin Padian, and Peter Ward, who read the entire manuscript in an earlier form and offered a wealth of useful suggestions. This book wouldn't exist without the intellectual freedom maintained by Marc Zabludoff, Paul Hoffman, and Robert Kunzig at *Discover*; the creative energies of illustrator Carl Buell; the efforts of my agent, Eric Simonoff; and the sustaining curiosity of my editor, Stephen Morrow.

The illustrations on pages 5 and 73 are partially adapted from W. Gilbert, *Developmental Biology*, 4th ed. (Sunderland, Mass.: Sinauer, 1994); on page 76 from C. Tabin, "The initiation of the limb bud: Growth factors, *Hox* genes, and retinoids," *Cell* 1995, 80:671–674; on page 78 from N. Shubin, C. Tabin, and S. Carroll, "Fossils, genes, and the evolution of animal limbs," *Nature* 1997, 388: 639–648; on page 81 from G. B. Müller and J. Streicher, "Ontogeny of the syndesmosis tibiofibularis and the evolution of the bird hindlimb: a caenogenetic feature triggers phenotypic novelty," *Anatomy and Embryology* 1989, 179:327–339; on page 92 from T. W. Pietsch and D. B. Grobecker, *Frogfishes of the World: Systematics, Zoogeography, and Behavioral Ecology* (Stanford: Stanford University Press, 1987); on page 185, F. E. Fish, "Association of propulsive mode with behavior in river otters *(Lutra canadensis)*," *Journal of Mammalogy* 1994, 75:989–997.

Triceratops, 151

Trilobites, 48

Trout, 101

Tulerpeton, 49–50, 72, 99, 106, 107, 243

Tuna, 46

Turing, Alan, 63–66

Turing patterns, 64–65, 72, 85, 98

Turtles, 59–60, 69, 109, 167, 197, 198, 199

Tyrannosaurus, 151, 182

Uhen, Mark, 177–78, 193, 204

Ulna, 55–56, 58, 67, 70, 79

Ungulates, 147, 153, 154, 156, 219, 243. *See also* specific animals

United States National History Museum (Washington, D.C.), 176–77

University of Arizona, 212

University of Austria, 81

University of Brussels, 213

University of California, 166

University of California at Berkeley, 66

University of California at Los Angeles, 133, 219

University of Chicago, 154

University of Hawaii, 130–31

University of Michigan, 159, 164, 188

University of North Carolina, 124

University of Pennsylvania, 40, 41

University of Puerto Rico, 216

University of Texas, 174

University of Toronto, 113–15

University of Utrecht, 188

University of Washington, 234

Urea, 46–48

Urine, 46–48

Van Valen, Leigh, 153–55, 162, 165, 206–207

Venjukoviamorphs, 149

Ventastega, 99, 105

Vertebrates: Archetype of, 20–21, 22,

57–59, 69, 146, 170; brains of, 220; embryos of, 73; first vertebrate, 5; homologies throughout anatomies of, 20, 22–23; primitive vertebrates, 28; sea squirt as relative of, 60; as tetrapods, 5–6. *See also* specific animals

Vestibular system, 87

Vestiges, 169–76

Victoria, Queen, 20

Von Baer, Karl, 59

Vrba, Elizabeth, 38

Wagner, Guenter, 80–84, 151

Walking of amniotes, 150

Warthogs, 153

Washington State, 209

Water compared to air, 6–7

Water rats, 183

Water shrews, 180

Watson, D. M. S., 61

Wells, Neil, 164

West Chester University, 179–80

Whales: *Aetiocetus*, 206; *Ambulocetus*, 194–97, 199–201, 203, 217, 228, 240; archaeocetes, 148, 197, 206–207, 209–10, 214, 215, 219, 223, 240; *Archaeomysticetus*, 207–208; Aristotle on, 144; baleen whales, 118, 119, 134, 205, 206–10, 213–15, 218, 223, 235; baseball vertebra of, 204; *Basilosaurus*, 139–44, 146–48, 166–70, 176–78, 193, 194, 202–207, 240; blowhole of, 209–10; body temperature of, 205; brains of, 219, 221, 223–26; cladogram of, 182, 218; compared with hippos, 212, 213, 218; compared with tetrapods, 228–30; *Dalanistes*, 200, 201; Darwin on, 144–45; *Dorudon*, 167, 168, 177–78, 203–205, 223, 229, 241; ears of, 130, 164–65, 190, 192, 199, 205, 209, 210, 229; and

Whales (*cont.*)
echolocation, 118, 126, 130, 205, 206, 209–10, 214–15; exaptations of, 118; family units of, 224–25; as filter feeders, 208–209, 214–15, 234–35; Flower's study of, 145–48, 218; fossils of, 137–44, 146, 159–69, 176–79, 186–97, 200–201, 209; *Gaviocetus*, 200; gene trees of, 211–19; general characteristics of, 117; Gingerich's study of, 159–69, 176–78, 179; head of, 126; hips of, 168, 177; and hunting for food, 118–19, 202, 207, 225; jaws of, 205, 206, 207, 208–209, 210, 214; kidneys of, 201, 202; legs of, 169, 170, 171, 173, 176–78, 192–96, 202, 204, 229; lifespan of, 224; mysticetes, 118, 146, 148, 203, 206–207, 210, 214, 218, 223, 242; odontocetes, 118–19, 126, 130, 146, 148, 203, 206–10, 213, 214, 218, 223–24, 242; origin and evolution of, 6, 118, 147–48, 152–58, 164–69, 177–78, 191–92, 194–226, 228–30; oxygen isotopes in, 201–202; *Pakicetus*, 164–68, 178, 179, 182, 186–92,
199–203, 213, 242; phylogeny of, 203; *Rodhocetus*, 195, 196, 197, 200; sounds made by, 126, 128, 209, 210, 224; spine of, 168, 177, 195–96, 202, 204; swimming of, 183–86, 199, 204, 205, 229; *Takracetus*, 200; teeth of, 137–41, 153–54, 165, 167, 192, 200, 206, 207, 208, 213, 214; Thewissen's study of, 186–96; toothed whales, 118, 119, 126, 128, 130, 205, 206, 208–10; unanswered questions about, 119
Wilhelm IV of Prussia, 143–44
Wolves, 145
Worms, 73
Wrist bones, 61, 62, 69, 70, 84, 85

Xiaoyuan Zhou, 194

Yale University, 38, 80, 165
Yapoks, 183, 184, 186, 199
Yellowtail snappers, 1–5

Zebra fish, 77, 119
Zeuglodon, 140
Zeuglodon Valley, 166–68, 193, 204
Zone of Polarizing Activity, 75, 76, 77